# Math Games

## Skill-Based Practice for Sixth Grade

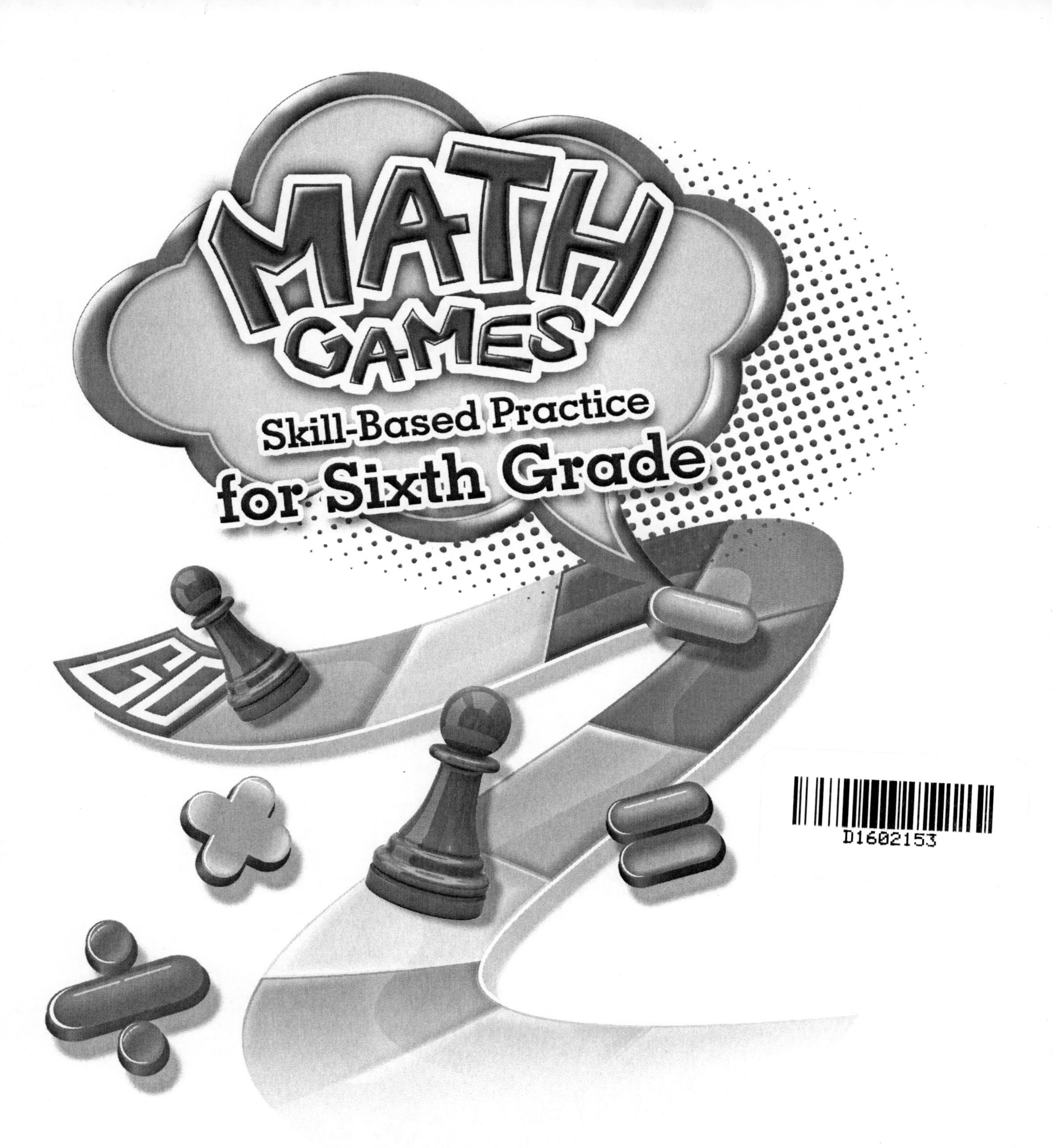





**Authors**

Ted H. Hull, Ed.D.

Ruth Harbin Miles, Ed.S.

Don S. Balka, Ph.D.

## Consultant

**Don W. Scheuer, Jr., M.S.Ed.**
*Mathematics Specialist*
The Haverford School (ret.)

## Publishing Credits

Robin Erickson, *Production Director;*
Lee Aucoin, *Creative Director;* Tim J. Bradley, *Illustration Manager;*
Sara Johnson, M.S.Ed., *Editorial Director;* Maribel Rendón, M.A.Ed., *Editor;*
Jennifer Viñas, *Editor;* Sara Sciuto, *Assistant Editor;* Grace Alba, *Designer;*
Corinne Burton, M.A.Ed., *Publisher*

## Image Credits

All images Shutterstock

## Standards



---

**Shell Education**
5301 Oceanus Drive
Huntington Beach, CA 92649-1030
http://www.shelleducation.com

**ISBN 978-1-4258-1293-5**

# Table of Contents

# Table of Contents *(cont.)*

# Importance of Games

Students learn from play. Play begins when we are infants and continues through adulthood. Games are motivational and educational (Hull, Harbin Miles, and Balka 2013; Burns 2009). They can assist and encourage students to operate as learning communities by requiring students to work together by following rules and being respectful. Games also foster students' thinking and reasoning since students formulate winning strategies. They provide much more sustained practices than do worksheets, and students are more motivated to be accurate. Worksheets may provide 20 to 30 opportunities for students to practice a skill, while games far exceed such prescribed practice opportunities. Lastly, games provide immediate feedback to students concerning their abilities.

Games must be part of the overall instructional approach that teachers use because successful learning requires active student engagement (Hull, Harbin Miles, and Balka 2013; National Research Council 2004), and games provide students with the motivation and interest to become highly engaged. Instructional routines need balance between concept development and skill development. They must also balance teacher-led and teacher-facilitated lessons. Students need time to work independently and collaboratively in order to assimilate information, and games can help support this.

When games are used appropriately, students also learn mathematical concepts.

## Mathematical Learning

Students must learn mathematics with understanding (NCTM 2000). Understanding means that students know the relationship between mathematical concepts and mathematical skills—mathematical procedures and algorithms work because of the underlying mathematical concepts. In addition, skill proficiency allows students to explore more rigorous mathematical concepts. From this relationship, it is clear that a balance between skill development and conceptual development must exist. There cannot be an emphasis of one over the other.

The National Council of Teachers of Mathematics (2000) and the National Research Council (2001) reinforce this idea. Both organizations state that learning mathematics requires both conceptual understanding and procedural fluency. This means that students need to practice procedures as well as develop their understanding of mathematical concepts in order to achieve success. The games presented in this book reinforce skill-based practice and support students' development of proficiency. These games can also be used as a springboard for discourse about mathematical concepts. The counterpart to this resource is *Math Games: Getting to the Core of Conceptual Understanding*, which builds students' conceptual understanding of mathematics through games.

# Importance of Games *(cont.)*

The *Common Core State Standards for Mathematics* (2010) advocate a balanced mathematics curriculum by focusing standards both on mathematical concepts and skills. This is also stressed in the Standards for Mathematical Practice, which discuss the process of "doing" mathematics and the habits of mind students need to possess in order to be successful.

The Standards for Mathematical Practice also focus on the activities that foster thinking and reasoning in which students need to be involved while learning mathematics. Games are an easy way to initiate students in the development of many of the practices. Each game clearly identifies a Common Core domain, a standard, and a skill, and allows students to practice them in a fun and meaningful way.

## Games vs. Worksheets

In all likelihood, many mathematics lessons are skill related and are taught and practiced through worksheets. Worksheets heavily dominate elementary mathematics instruction. They are not without value, but they often command too much time in instruction. While students need to practice skills and procedures, the way to practice these skills should be broadened.

Worksheets generally don't promote thinking and reasoning. They become so mechanical that students cease thinking. They are lulled into a feeling that completing is the goal. This sense of "just completing" is not what the Common Core Standards for Mathematical Practice mean when they encourage students to "persevere in solving problems."

Students need to be actively engaged in learning.

Students need to be actively engaged in learning. While worksheets do serve a limited purpose in skill practice, they also contain many potential difficulties. Problems that can occur include the following:

- **Worksheets are often completed in isolation,** meaning that students who are performing a skill incorrectly most likely practice the skill incorrectly for the entire worksheet. The misunderstanding may not be immediately discovered, and in fact, will most likely not be discovered for several days!
- **Worksheets are often boring to students.** Learning a skill correctly is not the students' goal. Their goal becomes to finish the worksheet. As a result, careless errors are often made, and again, these errors may not be immediately discovered or corrected.

 

# Importance of Games *(cont.)*

- ➔ **Worksheets are often viewed as a form of subtle punishment.** While perhaps not obvious, the perceived punishment is there. Students who have mastered the skill and can complete the worksheet correctly are frequently "rewarded" for their efforts with another worksheet while they wait for their classmates to finish. At the same time, students who have not mastered the skill and do not finish the worksheet on time are "rewarded" with the requirement to take the worksheet home to complete, or they finish during another portion of the day, often recess or lunch.
- ➔ **Worksheets provide little motivation to learn a skill correctly.** There is no immediate correction for mistakes, and often, students do not really care if a mistake is made. When a game is involved, students want and need to get correct answers.

The *Common Core State Standards for Mathematics*, including the Standards for Mathematical Practice, demand this approach change. These are the reasons teachers and teacher leaders must consciously support the idea of using games to support skill development in mathematics.

# How to Use This Book

There are many ways to effectively utilize this book. Teachers, mathematics leaders, and parents may use this book to engage students in fun, meaningful, practical mathematics learning. These games can be used as a way to help students maintain skill proficiency or remind them of particular skills prior to a critical concept lesson. These games may also be useful during tutorial sessions, or during class when students have completed their work.

## Games at Home

Parents may use these games to work with their child to learn important skills. The games also provide easier ways for parents to interest their child in learning mathematics rather than simply memorizing facts. In many cases, their child is more interested in listening to explanations than correcting their errors.

Parents want to help their children succeed in school, yet they may dread the frequently unpleasant encounters created by completing mathematics worksheets at home. Families can easily use the games in this book by assuming the role of one of the players. At other times, parents provide support and encouragement as their child engages in the game. In either situation, parents are able to work with their children in a way that is fun, educational, and informative.

## Games in the Classroom

During game play, teachers are provided excellent opportunities to assess students' abilities and current skill development. Students are normally doing their best and drawing upon their current understanding and ability to play the games, so teachers see an accurate picture of student learning. Some monitoring ideas for teacher assessment include:

- Move about the room listening and observing
- Ask student pairs to explain what they are doing
- Ask the entire class about the game procedures after play
- Play the game against the class
- Draw a small group of students together for closer supervision
- Gather game sheets to analyze students' proficiencies

Ongoing formative assessment and timely intervention are cornerstones of effective classroom instruction. Teachers need to use every available opportunity to make student thinking visible and to respond wisely to what students' visible thinking reveals. Games are an invaluable instructional tool that teachers need to effectively use.

# How to Use This Book *(cont.)*

Students are able to work collaboratively during game play, thus promoting student discourse and deeper learning. The games can also be used to reduce the amount of time students spend completing worksheets.

Each game in this book is based upon a common format. This format is designed to assist teachers in understanding how the game activities are played and which standards and mathematical skills students will be practicing.

**Domain**
The domain that students will practice is noted at the beginning of each lesson. Each of the five domains addressed in this series has its own icon.

**Standards**
One or more *Common Core State Standards* will state the specific skills that students will practice during game play.

**Number of Players**
The number of players varies for each game. Some may include whole-group game play, while others may call for different-size groups.

**Materials**
A materials list is provided for each game to notify the teacher what to have available in order to play the games.

**Get Prepared!**
Everything a teacher needs to be prepared for game play is noted in the Get Prepared! section. This includes how many copies are needed as well as other tasks that need to be completed with the materials.

**Game Directions**
The directions allow for step-by-step guidance on how to easily implement each game.

All game resources can be found on the **Digital Resource CD**. (For a complete list of the files, see pages 135–136.)

# How to Use This Book *(cont.)*

Many games include materials such as game boards, activity cards, and score cards. You may wish to laminate materials for durability.

**Game Boards**

Some game boards spread across multiple book pages in order to make them larger for game play. When this is the case, cut out each part of the game board and tape them together. Once you cut them apart and tape them together, you may wish to glue them to a large sheet of construction paper and laminate them for durability.

**Activity Cards**

Some games include activity cards. Once you cut them apart, you may wish to laminate them for durability.

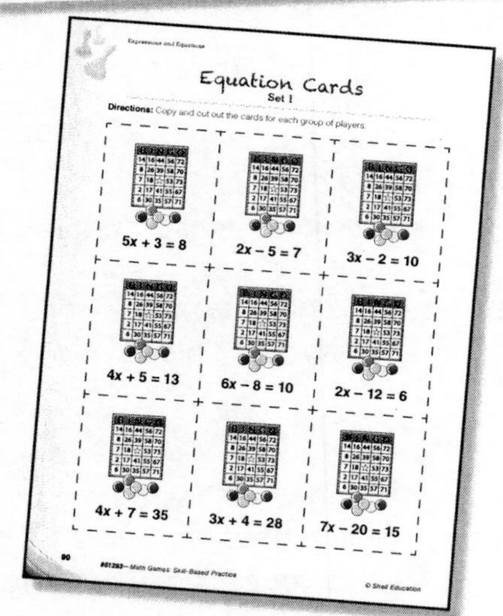

Equation Cards

Set 1

**Directions:** Copy and cut out the cards for each group of players.

| | | |
|---|---|---|
| $5x + 3 = 8$ | $2x - 5 = 7$ | $3x - 2 = 10$ |
| $4x + 5 = 13$ | $6x - 8 = 10$ | $2x - 12 = 6$ |
| $4x + 7 = 35$ | $3x + 4 = 28$ | $7x - 20 = 15$ |

# Correlation to the Standards

Shell Education is committed to producing educational materials that are research and standards based. In this effort, we have correlated all of our products to the academic standards of all 50 United States, the District of Columbia, the Department of Defense Dependent Schools, and all Canadian provinces.

## How to Find Standards Correlations

To print a customized correlation report of this product for your state, visit our website at **http://www.shelleducation.com** and follow the on-screen directions. If you require assistance in printing correlation reports, please contact Customer Service at 1-877-777-3450.

## Purpose and Intent of Standards

Legislation mandates that all states adopt academic standards that identify the skills students will learn in kindergarten through grade twelve. Many states also have standards for Pre–K. This same legislation sets requirements to ensure the standards are detailed and comprehensive.

Standards are designed to focus instruction and guide adoption of curricula. Standards are statements that describe the criteria necessary for students to meet specific academic goals. They define the knowledge, skills, and content students should acquire at each level. Standards are also used to develop standardized tests to evaluate students' academic progress. Teachers are required to demonstrate how their lessons meet state standards. State standards are used in the development of all of our products, so educators can be assured they meet the academic requirements of each state.

## Common Core State Standards

Many games in this book are aligned to the Common Core State Standards. The standards support the objectives presented throughout the lessons and are provided on the Digital Resource CD (standards.pdf).

## TESOL and WIDA Standards

The lessons in this book promote English language development for English language learners. The standards listed on the Digital Resource CD (standards.pdf) support the language objectives presented throughout the lessons.

# Standards Correlation Chart

| Standard | Game(s) |
|---|---|
| **6.RP.1** Understand the concept of a ratio and use ratio language to describe a ratio relationship between two quantities. | Climb the Mountain (p.16); Rocking Ratios (p. 26) |
| **6.RP.3** Use ratio and rate reasoning to solve real-world and mathematical problems, e.g., by reasoning about tables of equivalent ratios, tape diagrams, double number line diagrams, or equations. | Roller Coaster Proportions (p. 20) |
| **6.NS.1** Interpret and compute quotients of fractions, and solve word problems involving division of fractions by fractions, e.g., by using visual fraction models and equations to represent the problem. | Between (p. 34); Quotient Corral (p. 55) |
| **6.NS.3** Fluently add, subtract, multiply, and divide multi-digit decimals using the standard algorithm for each operation. | Dunking Decimals (p. 38) |
| **6.NS.4** Find the greatest common factor of two whole numbers less than or equal to 100 and the least common multiple of two whole numbers less than or equal to 12. | Mystery Multiples (p. 42); GCF for the Win (p. 47) |
| **6.NS.6a** Recognize opposite signs of numbers as indicating locations on opposite sides of 0 on the number line; recognize that the opposite of the opposite of a number is the number itself, e.g., $-(-3) = 3$, and that 0 is its own opposite. | Zero to Hero (p. 58) |
| **6.NS.6c** Find and position integers and other rational numbers on a horizontal or vertical number line diagram; find and position pairs of integers and other rational numbers on a coordinate plane. | Rational Order (p. 60) |
| **6.EE.2** Write, read, and evaluate expressions in which letters stand for numbers. | Express Yourself with Expressions (p. 67) |
| **6.EE.2a** Write expressions that record operations with numbers and with letters standing for numbers. | Express Yourself with Expressions (p. 67) |

# Standards Correlation Chart *(cont.)*

| Standard | Game(s) |
|---|---|
| **6.EE.2c** Evaluate expressions at specific values of their variables. Include expressions that arise from formulas used in real-world problems. Perform arithmetic operations, including those involving whole number exponents, in the conventional order when there are no parentheses to specify a particular order (Order of Operations). | Most Valuable 7 (p. 75) |
| **6.EE.3** Apply the properties of operations to generate equivalent expressions. | Distribution Concentration (p. 82) |
| **6.EE.5** Understand solving an equation or inequality as a process of answering a question: which values from a specified set, if any, make the equation or inequality true? Use substitution to determine whether a given number in a specified set makes an equation or inequality true. | Equation Bingo (p. 88); True or False: Inequalities (p. 101) |
| **6.G.2** Find the volume of a right rectangular prism with fractional edge lengths by packing it with unit cubes of the appropriate unit fraction edge lengths, and show that the volume is the same as would be found by multiplying the edge lengths of the prism. Apply the formulas $V = l\ w\ h$ and $V = b\ h$ to find volumes of right rectangular prisms with fractional edge lengths in the context of solving real-world and mathematical problems. Represent three-dimensional figures using nets made up of rectangles and triangles, and use the nets to find the surface area of these figures. Apply these techniques in the context of solving real-world and mathematical problems. | High Velocity Volume (p. 106) |
| **6.G.3** Draw polygons in the coordinate plane given coordinates for the vertices; use coordinates to find the length of a side joining points with the same first coordinate or the same second coordinate. Apply these techniques in the context of solving real-world and mathematical problems. | Triangles (p. 111) |
| **6.SP.4** Display numerical data in plots on a number line, including dot plots, histograms, and box plots. | Mean Wins (p. 114) |

# Standards Correlation Chart *(cont.)*

| Standard | Game(s) |
|---|---|
| **6.SP.5c** Giving quantitative measures of center (median and/or mean) and variability (interquartile range and/or mean absolute deviation), as well as describing any overall pattern and any striking deviations from the overall pattern with reference to the context in which the data were gathered. | Dive Into Distributions (p. 118); Statistics Strikeout (p. 122) |

# About the Authors

**Ted H. Hull, Ed.D.,** served in public education for 32 years as a mathematics teacher, a K–12 mathematics coordinator, a school principal, director of curriculum and instruction, and project director for the Charles A. Dana Center at the University of Texas in Austin. While at the University of Texas, he directed the research project "Transforming Schools: Moving from Low-Achieving to High Performing Learning Communities." After retiring, Ted opened LCM: Leadership • Coaching • Mathematics with his coauthors and colleagues. Ted has coauthored numerous books addressing mathematics improvement and has served as the Regional Director for the National Council of Supervisors of Mathematics (NCSM).

**Ruth Harbin Miles, Ed.S.,** currently coaches inner-city, rural, and suburban mathematics teachers and serves on the Board of Directors for the National Council of Teachers of Mathematics, the National Council of Supervisors of Mathematics and Virginia's Council of Mathematics Teachers. Her professional experiences include coordinating the K–12 Mathematics Department for Olathe, Kansas Schools and adjunct teaching for Mary Baldwin College and James Madison University in Virginia. A coauthor of four books on transforming teacher practice through team leadership, mathematics coaching, and visible student thinking and co-owner of Happy Mountain Learning, Ruth's specialty and passion include developing teachers' content knowledge and strategies for engaging students to achieve high standards in mathematics.

**Don S. Balka, Ph.D.,** a former middle school and high school mathematics teacher, is Professor Emeritus in the Mathematics Department at Saint Mary's College in Notre Dame, Indiana. Don has presented at over 2,000 workshops, conferences, and in-service trainings throughout the United States and has authored or coauthored over 30 books on mathematics improvement. Don has served as director for the National Council of Teachers of Mathematics, the National Council of Supervisors of Mathematics, TODOS: Mathematics for All, and the School Science and Mathematics Association. He is currently president of TODOS and past president of the School Science and Mathematics Association.

# Climb the Mountain

## Domain **A:B**

Ratios and Proportional Relationships

## Standard

Understand the concept of a ratio and use ratio language to describe a ratio relationship between two quantities.

## Number of Players

2 to 3 Players

## Materials

- *Climb the Mountain Game Sheet* (page 18)
- *Climb the Mountain Game Markers* (page 19)
- regular decks of cards
- number cubes

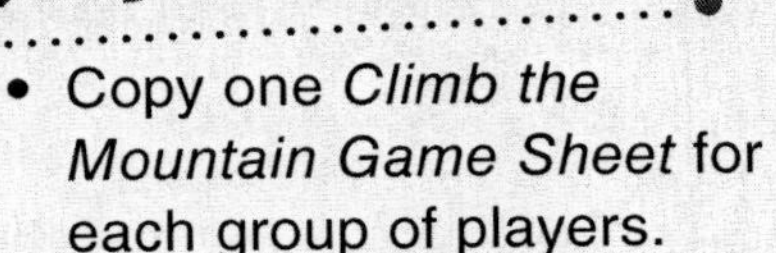


- Copy one *Climb the Mountain Game Sheet* for each group of players.
- Copy and cut out the *Climb the Mountain Game Markers* for the class.
- Collect a deck of cards and a number cube for each group of players.

## Game Directions

1. Distribute materials to players and have each player select one of the *Climb the Mountain Game Markers*.
2. Players take turns rolling a number cube. The player who rolls the highest number is Player 1.
3. Player 1 deals eight cards to each player and places the remaining deck facedown.
4. For face cards, the Jack has a value of 12, the Queen has a value of 15, and the King has a value of 16. The Ace has a value of 1.

 

# Climb the Mountain *(cont.)*

5. The player to the right of Player 1 selects two cards from his or her hand to create a fraction equal to or equivalent to one on the *Climb the Mountain Game Sheet*.

   The player displays the fraction and covers the matching space with a game marker. A marker can cover only one fraction.

6. The two cards are placed facedown in a discard pile and two new cards are drawn.

7. A player who cannot create a fraction on the sheet may discard two cards and draw two new cards, but must wait to play on the next turn.

8. The game ends when all cards are drawn.

9. The player with the most markers on the game sheet wins.

Name: ______________________________ Date: ________________

# Climb the Mountain

## Game Sheet

**Directions:** Drawing two cards from the deck, create a proper fraction or 1. Then place a marker on the fraction or equivalent fraction below.

# Climb the Mountain

## Game Markers

**Directions:** Copy and cut out a game marker for each player.

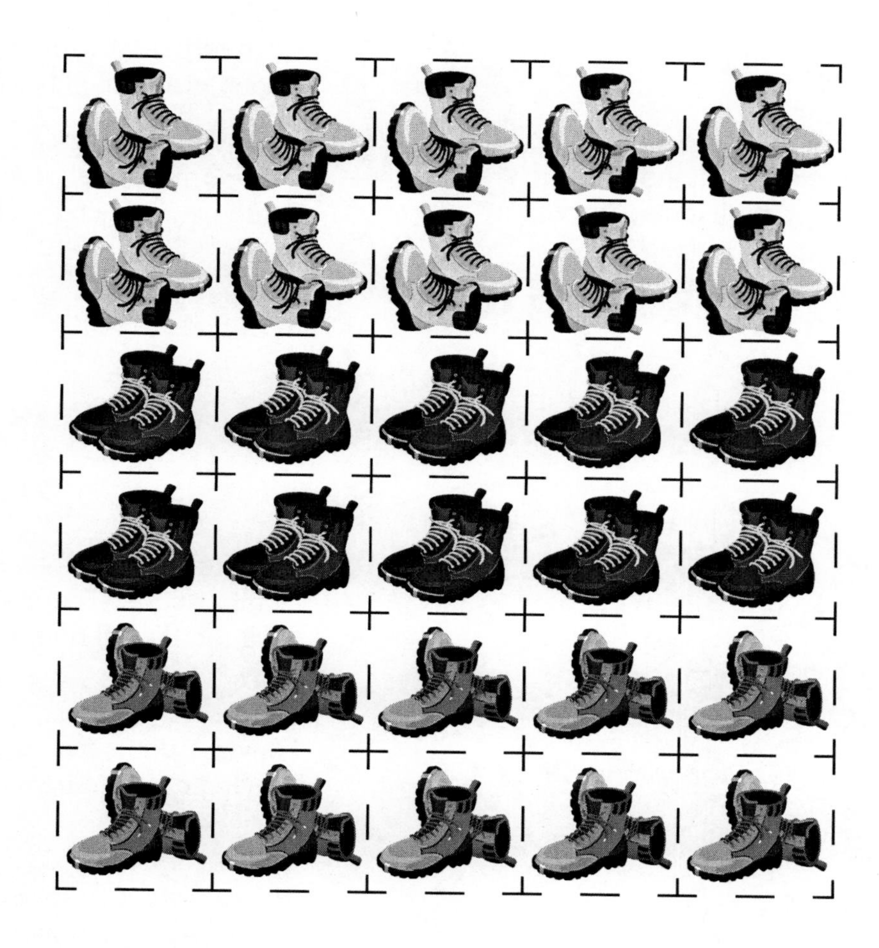

# Roller Coaster Proportions

## Domain A:B

Ratios and Proportional Relationships

## Standard

Use ratio and rate reasoning to solve real-world and mathematical problems, e.g., by reasoning about tables of equivalent ratios, tape diagrams, double number line diagrams, or equations.

## Number of Players

2 to 3 Players

## Materials

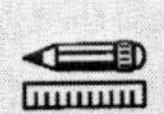

- *Roller Coaster Proportions Cards* (pages 22–24)
- *Roller Coaster Proportions Recording Sheet* (page 25)
- *Roller Coaster Proportions Answer Key* (page 131)
- regular decks of cards without face cards
- number cubes

## GET PREPARED!

- Copy and cut out a set of *Roller Coaster Proportions Cards* for each group of players.
- Copy two *Roller Coaster Proportions Recording Sheets* for each player.

## Game Directions

1. Distribute materials to players.
2. Players take turns rolling a number cube. The player who rolls the lowest number is Player 1.
3. Player 1 deals five regular playing cards to each player.
4. The set of *Roller Coaster Proportions Cards* is placed facedown on the play area.
5. Player 1 draws a card from the *Roller Coaster Proportions Cards* deck and places it on the playing area. Player 1 solves the problem on the *Roller Coaster Proportions Recording Sheet* and announces the solution to the other players. Answers can be checked using the *Roller Coaster Proportions Answer Key*.

# Roller Coaster Proportions *(cont.)*

6. Players lay down all cards in their hands that can be added or multiplied to get the solution. For example, the answer to the proportion on card #3 is 9. If a player has a "9," he or she can lay it down. If a player has "ace," "2," and "6," all three cards can be played since the sum is 9.

7. Players 2 and 3 repeat step 5.

8. Again, players lay down all cards in their hands that can be added or multiplied to get the solution.

9. The winner is the player who first gets rid of all five cards, or the player with the fewest cards if all *Roller Coaster Proportions Cards* are used.

# Roller Coaster Proportions

## Cards

**Directions:** Copy and cut out cards for each group of players.

| | | |
|---|---|---|
| 1 cup of flour for 3 cups of sugar<br>? cups of flour for 6 cups of sugar<br>1 | $5 for 4 hamburgers<br>$10 for ? hamburgers<br>2 | 1 robin with 2 legs<br>? robins with 18 legs<br>3 |
| $12 for 3 balls<br>$24 for ? balls<br>4 | 2 tsp. of salt to 4 cups of water<br>? tsp. of salt to 10 cups of water<br>5 | 2 quarts to 7 gallons<br>? quarts to 28 gallons<br>6 |
| 3 meters to 6 centimeters<br>5 meters to ? centimeters<br>7 | 2 pencils to 5 pens<br>? pencils to 10 pens<br>8 | 120 miles in 15 hours<br>56 miles in ? hours<br>9 |
| 18 girls to 36 boys<br>? girl(s) to 2 boys<br>10 | 8 orange balls to 24 green balls<br>1 orange ball to ? green balls<br>11 | 3 liters for 48 square meters<br>? liters for 144 square meters<br>12 |

# Roller Coaster Proportions

## Cards *(cont.)*

| | | |
|---|---|---|
| 2 teachers per 24 students<br><br>? teachers per 72 students<br><br>13 | 3 batteries for $4.80<br><br>? batteries for $14.40<br><br>14 | 60 chicken wings for 20 students<br><br>27 chicken wings for ? students<br><br>15 |
| 3 T-shirts for $15<br><br>? T-shirts for $10<br><br>16 | 4 blocks in 3 minutes<br><br>8 blocks in ? minutes<br><br>17 | 12 apples to 15 peaches<br><br>4 apples to ? peaches<br><br>18 |
| $42 per 3 hours<br><br>$28 per ? hours<br><br>19 | 2 eggs to 5 cups of milk<br><br>? eggs to 10 cups of milk<br><br>20 | 3 minutes to 21 hours<br><br>1 minute to ? hours<br><br>21 |
| 30 TVs for 20 houses<br><br>12 TVs for ? houses<br><br>22 | 210 miles on 30 gallons<br><br>7 miles on ? gallons<br><br>23 | 14 yards for 2 costumes<br><br>49 yards for ? costumes<br><br>24 |

# Roller Coaster Proportions

## Cards *(cont.)*

| | | |
|---|---|---|
| 1025 sq. ft. per 5 pounds<br>2050 sq. ft. per ? pounds<br>25 | 5 cups of flour to 1 cup of milk<br>30 cups of flour to ? cups of milk<br>26 | $180 for family of 6<br>$90 for family of ?<br>27 |
| 72 beats per minute<br>360 beats per ? minutes<br>28 | 1 gallon to 16 cups<br>? gallons to 64 cups<br>29 | $50 for 5 books<br>$10 for ? books<br>30 |
| 5 oranges for $2.25<br>? oranges for $4.50<br>31 | 5 dogs to 15 cats<br>? dogs to 27 cats<br>32 | 22 red blocks to 16 yellow blocks<br>11 red blocks to ? yellow blocks<br>33 |
| 14 inches to 35 feet<br>2 inches to ? feet<br>34 | 3 pounds for $18<br>1 pound for $?<br>35 | 35 black to 21 blue<br>5 black to ? blue<br>36 |
| $80 in 2 weeks<br>$280 in ? weeks<br>37 | 5 gallons for 160 miles<br>? gallons for 128 miles<br>38 | 6 red shirts to 12 blue shirts<br>? red shirts to 2 blue shirts<br>39 |

Name:______________________________ Date:________________

# Roller Coaster Proportions

## Recording Sheet

**Directions:** Use the recording sheet to solve the problems.

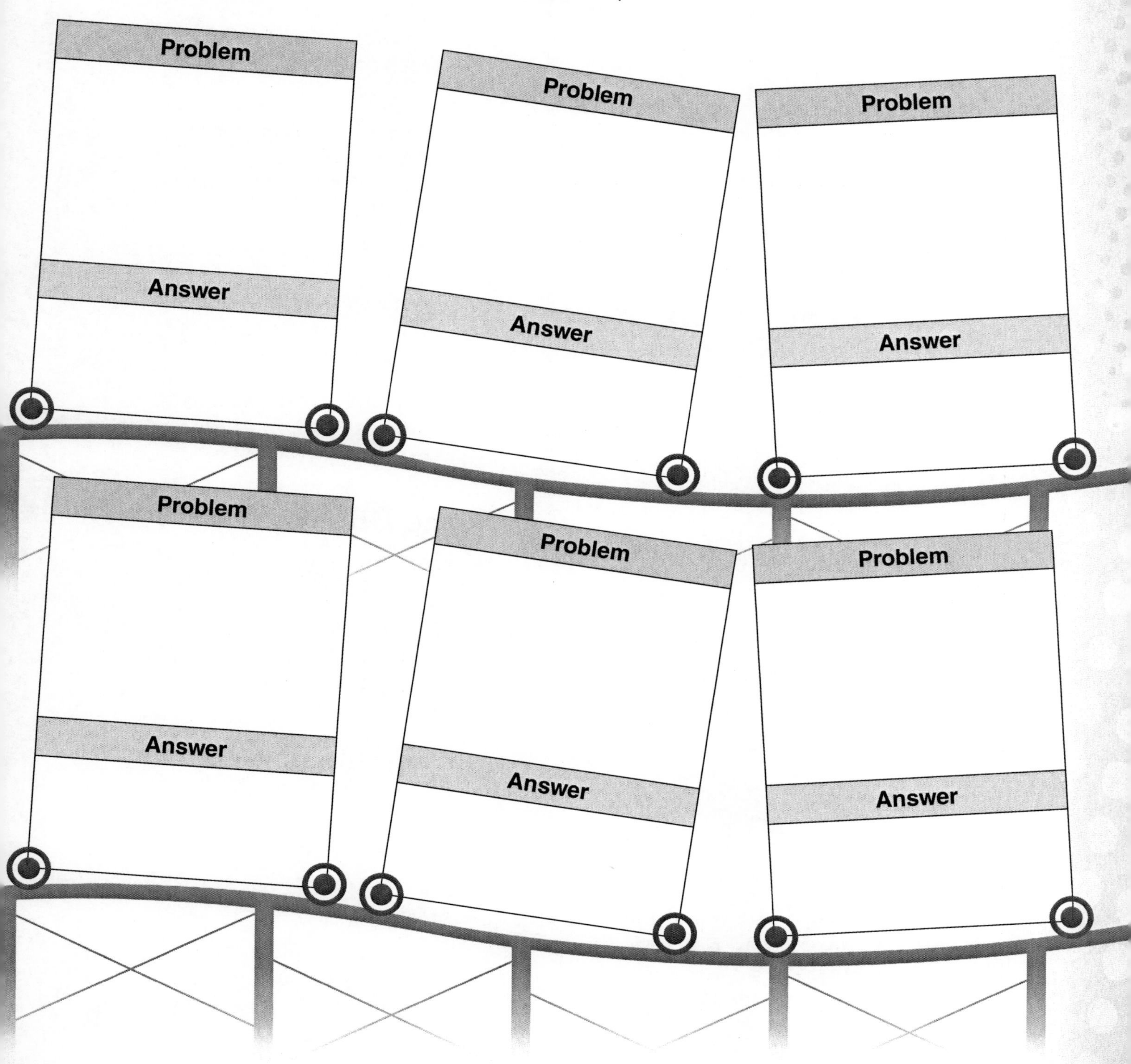

# Rocking Ratios

## Domain A:B

Ratios and Proportional Relationships

## Standard

Understand the concept of a ratio and use ratio language to describe a ratio relationship between two quantities.

## Number of Players

2 Players

## Materials

- *Rocking Ratios 1–100 Sheet* (page 28)
- *Rocking Ratios Cards* (pages 29–32)
- *Rocking Ratios Recording Sheet* (page 33)
- *Rocking Ratios Cards Answer Key* (page 132)

## GET PREPARED

- Copy and cut out a set of *Rocking Ratios Cards* for each pair of players.
- Copy one *Rocking Ratios 1–100 Sheet* and one *Rocking Ratios Recording Sheet* for each pair of players.
- Prior to the game, review the ratio cards and remove cards containing content students have not yet been taught.

## Game Directions

1. Distribute materials to players.
2. Players place the cards facedown. Players take turns turning over the *Rocking Ratios Cards*.

# Rocking Ratios *(cont.)*

3. Players work together to find the ratio indicated on the card by determining from the *Rocking Ratios 1–100 Sheet* the quantity of numbers in each specified set and record answers on the *Rocking Ratios Recording Sheet*. For example, there are 25 prime numbers and 74 composite numbers on the *Rocking Ratios 1–100 Sheet*. Therefore, the ratio of prime numbers to composite numbers is 25/74. Recall that 1 is not prime or composite.

4. Depending on what has been taught, players can simplify the ratios.

5. The winner is the pair of players that first correctly finds all the ratios.

6. Have pairs of players report the ratios found for each card. Answers can be checked using the *Rocking Ratios Cards Answer Key*.

# Rocking Ratios
## 1–100 Sheet

**Directions:** Use this sheet to find the quantity of numbers in each specified set.

| | | | | | | | | | |
|---|---|---|---|---|---|---|---|---|---|
| 1 | 2 | 3 | 4 | 5 | 6 | 7 | 8 | 9 | 10 |
| 11 | 12 | 13 | 14 | 15 | 16 | 17 | 18 | 19 | 20 |
| 21 | 22 | 23 | 24 | 25 | 26 | 27 | 28 | 29 | 30 |
| 31 | 32 | 33 | 34 | 35 | 36 | 37 | 38 | 39 | 40 |
| 41 | 42 | 43 | 44 | 45 | 46 | 47 | 48 | 49 | 50 |
| 51 | 52 | 53 | 54 | 55 | 56 | 57 | 58 | 59 | 60 |
| 61 | 62 | 63 | 64 | 65 | 66 | 67 | 68 | 69 | 70 |
| 71 | 72 | 73 | 74 | 75 | 76 | 77 | 78 | 79 | 80 |
| 81 | 82 | 83 | 84 | 85 | 86 | 87 | 88 | 89 | 90 |
| 91 | 92 | 93 | 94 | 95 | 96 | 97 | 98 | 99 | 100 |

# Rocking Ratios

## Cards

**Directions:** Copy and cut out cards for each pair of students.

# Rocking Ratios

## Cards *(cont.)*

**Ratio of Multiples of 10 to Multiples of 5**

10

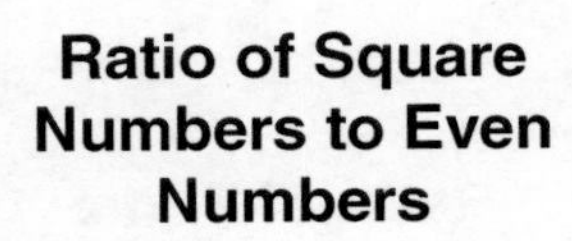

**Ratio of Square Numbers to Even Numbers**

11

**Ratio of Prime Numbers to Odd Numbers**

12

**Ratio of Numbers Greater Than 60 to Numbers Less Than 21**

13

**Ratio of Numbers Less Than 40 to Numbers Greater Than 49**

14

**Ratio of Numbers Greater Than 70 to Numbers Greater Than 60**

15

**Ratio of Multiples of 2 to Multiples of 4**

16

**Ratio of Multiples of 7 to Multiples of 8**

17

**Ratio of Numbers Containing a 6 to Numbers Ending with 2**

18

# Rocking Ratios

## Cards *(cont.)*

**Ratio of Numbers Containing a 9 to Numbers Ending with 3**

19

**Ratio of Square Numbers to Odd Numbers**

20

**Ratio of Triangular Numbers to Odd Numbers**

21

**Ratio of Numbers Whose Digits Sum Is 3 to Numbers Whose Digits Sum Is 8**

22

**Ratio of Numbers Whose Digits Sum Is 12 to Numbers Whose Digits Sum Is 14**

23

**Ratio of Numbers Whose Digits Sum Is 5 to Numbers Whose Digits Sum Is 9**

24

**Ratio of Multiples of 11 to Multiples of 13**

25

**Ratio of Multiples of 25 to Multiples of 20**

26

**Ratio of Multiples of 40 to Multiples of 30**

27

# Rocking Ratios

## Cards *(cont.)*

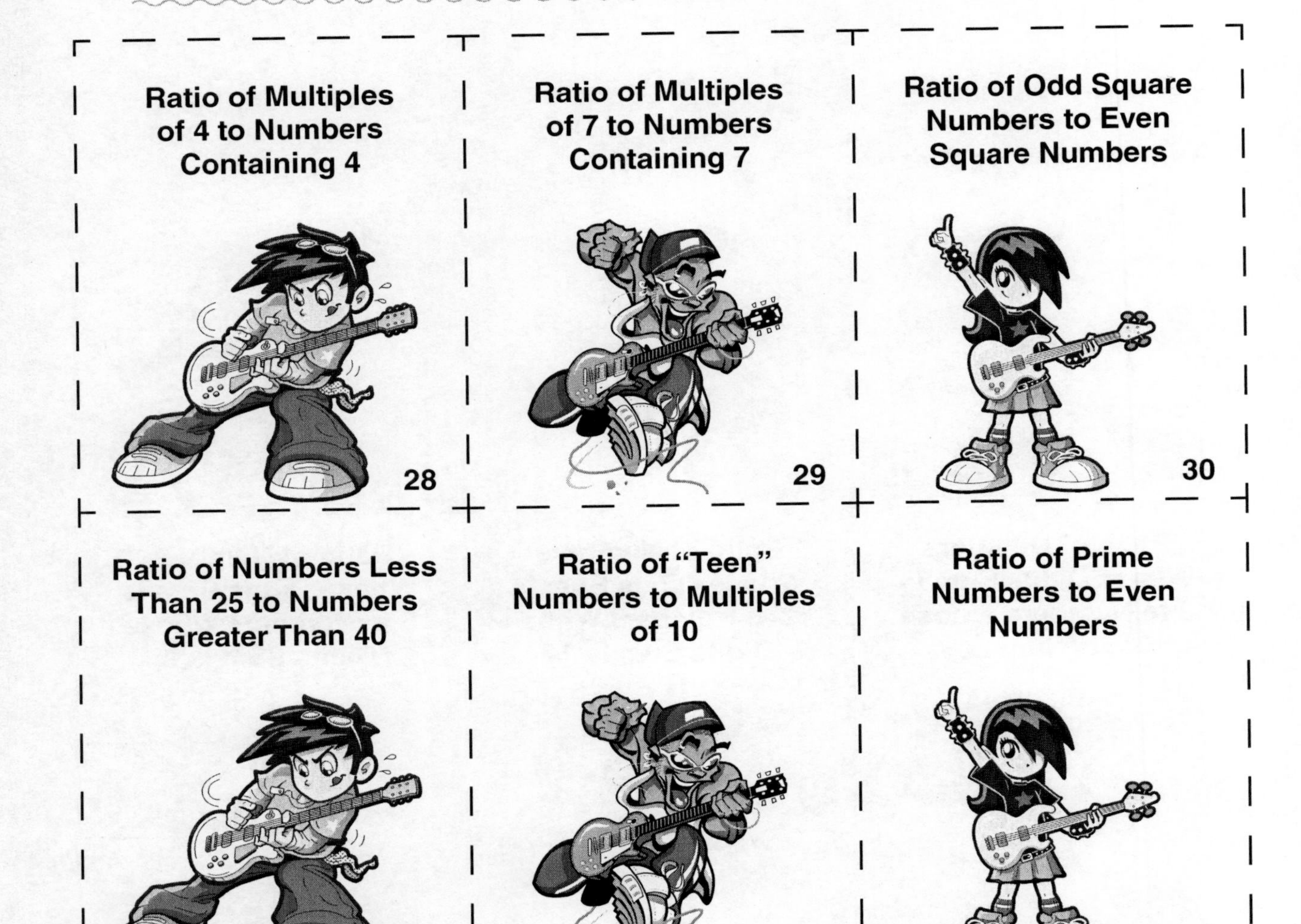


 

Name:________________________________ Date:__________________

# Rocking Ratios
## Recording Sheet

**Directions:** Record your answer for each ratio card on the recording sheet.

| Card Number | Answer | Card Number | Answer | Card Number | Answer |
|---|---|---|---|---|---|
| 1 | | 12 | | 23 | |
| 2 | | 13 | | 24 | |
| 3 | | 14 | | 25 | |
| 4 | | 15 | | 26 | |
| 5 | | 16 | | 27 | |
| 6 | | 17 | | 28 | |
| 7 | | 18 | | 29 | |
| 8 | | 19 | | 30 | |
| 9 | | 20 | | 31 | |
| 10 | | 21 | | 32 | |
| 11 | | 22 | | 33 | |

# Between

## Domain

The Number System

## Standard

Interpret and compute quotients of fractions, and solve word problems involving division of fractions by fractions.

## Number of Players

2 Players

## Materials

- *Between Game Sheet* (page 36)
- *Between Game Sheet Markers* (page 37)
- colored paper
- number cubes
- scratch paper
- calculators (to check answers)

## GET PREPARED!

- Copy a *Between Game Sheet* for each pair of players.
- Copy and cut out the *Between Game Sheet Markers* for each group of players using two different sheets of colored paper (one color for each player), so each player has eight markers.
- Collect a number cube and scratch paper for each pair of players.

## Game Directions

1. Distribute materials to players.
2. Players take turns rolling a number cube. The player who rolls the higher number is Player 1.
3. Player 1 selects two fractions at the bottom of the sheet and divides. For example, if Player 1 selects $\frac{1}{4}$ and $\frac{1}{3}$, he or she may find the quotient $\frac{1}{4} \div \frac{1}{3} = \frac{3}{4}$ or the quotient $\frac{1}{3} \div \frac{1}{4} = \frac{4}{3}$. Only one quotient can be used.

# Between *(cont.)*

4. Player 1 places one of his or her *Between Game Sheet Markers* in the square indicating the correct location of the quotient. In the example above, the first quotient would appear in the square marked $\frac{1}{2} < \frac{p}{q} \leq \frac{3}{4}$. The second quotient would appear in the square marked $\frac{5}{4} < \frac{p}{q} \leq \frac{3}{2}$. Players may check their answers using a calculator, as needed.

5. Player 2 repeats steps 3 and 4.

6. The first player to get three markers in a row horizontally, vertically, or diagonally is the winner.

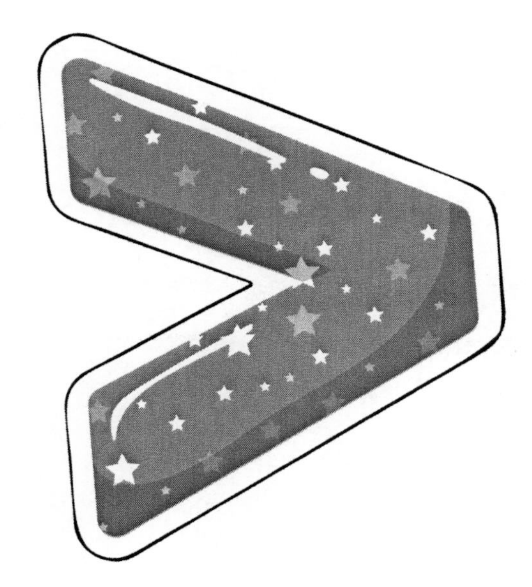

Name: ______________________________ Date: ______________

# Between
## Game Sheet

**Directions:** Select two fractions from the bottom of this sheet. Divide the fractions. Place a marker on the square that shows the location of the quotient. The first player to get three in a row horizontally, vertically, or diagonally is the winner.

| | | | |
|---|---|---|---|
| $0 < \frac{p}{q} \leq \frac{1}{8}$ | $\frac{1}{8} < \frac{p}{q} \leq \frac{1}{4}$ | $\frac{1}{4} < \frac{p}{q} \leq \frac{1}{2}$ | $\frac{1}{2} < \frac{p}{q} \leq \frac{3}{4}$ |
| $\frac{3}{4} < \frac{p}{q} \leq \frac{7}{8}$ | $\frac{1}{8} < \frac{p}{q} \leq \frac{1}{4}$ | $1 < \frac{p}{q} \leq \frac{9}{8}$ | $\frac{9}{8} < \frac{p}{q} \leq \frac{5}{4}$ |
| $\frac{5}{4} < \frac{p}{q} \leq \frac{3}{2}$ | $\frac{7}{8} < \frac{p}{q} \leq 1$ | $\frac{7}{4} < \frac{p}{q} \leq 2$ | $2 < \frac{p}{q} \leq \frac{9}{4}$ |
| $\frac{9}{4} < \frac{p}{q} \leq \frac{11}{4}$ | $\frac{3}{2} < \frac{p}{q} \leq \frac{7}{4}$ | $\frac{13}{4} < \frac{p}{q} \leq 4$ | $4 < \frac{p}{q} \leq 6$ |

| | | | | | | | |
|---|---|---|---|---|---|---|---|
| $\frac{1}{8}$ | $\frac{1}{6}$ | $\frac{1}{5}$ | $\frac{1}{4}$ | $\frac{1}{3}$ | $\frac{1}{2}$ | $\frac{2}{3}$ | $\frac{3}{4}$ |

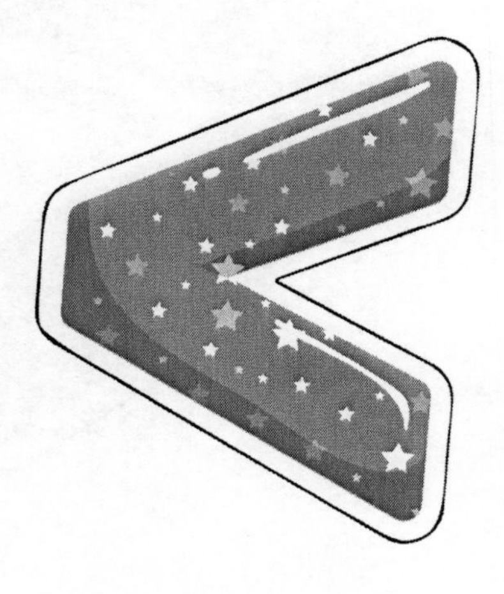

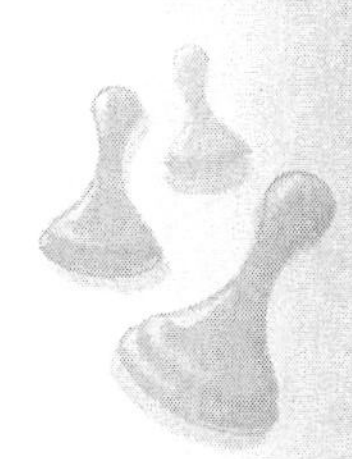

# Between

## Game Sheet Markers

**Directions:** Copy the four sets of markers on two different colored sheets of paper and cut out the markers. Each player should have a different colored set of eight markers.

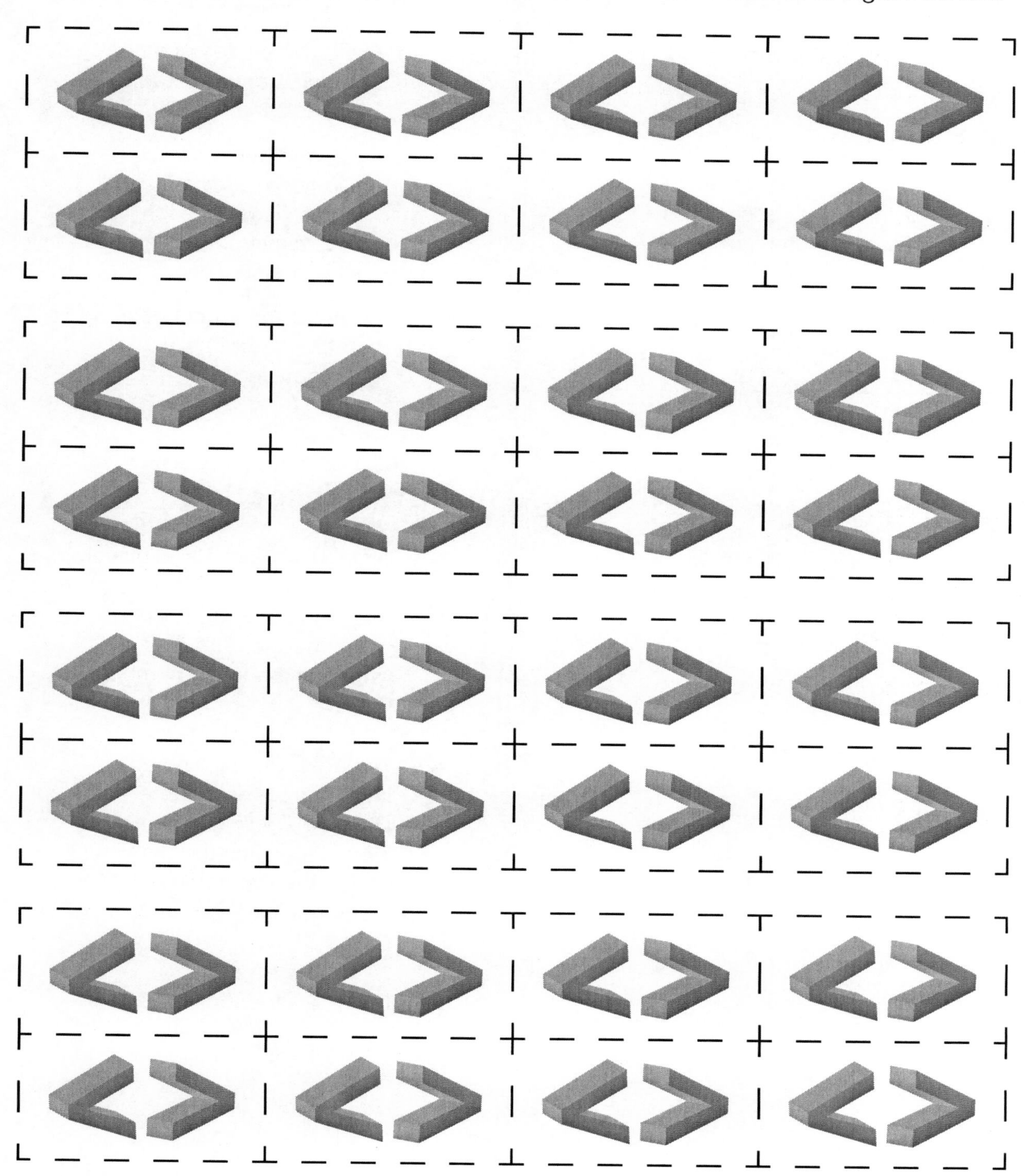

# Dunking Decimals

## Domain

The Number System

## Standard

Fluently add, subtract, multiply, and divide multi-digit decimals using the standard algorithm for each operation.

## Number of Players

2 Players

## Materials

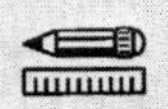

- *Dunking Decimals Game Sheet* (page 40)
- *Dunking Decimals Number Tiles* (page 41)
- paper bag
- number cube
- scratch paper

## GET PREPARED!

- Make two copies of the *Dunking Decimals Game Sheet* for each player.
- Copy and cut out a set of *Dunking Decimals Number Tiles* for each pair of players. Place the tiles in a paper bag.
- Collect a number cube and scratch paper for each pair of players.

## Game Directions

1. Distribute materials to players.
2. Players take turns rolling a number cube. The player who rolls the lower number is Player 1.
3. To begin, Player 1 reaches into the bag, draws one of the *Dunking Decimals Number Tiles* and writes the digit in one of the five squares on his or her *Dunking Decimals Game Sheet*.
4. Player 2 repeats step 3.

 

# Dunking Decimals *(cont.)*

5. Players take turns drawing a digit from the bag until all tiles have been drawn and all squares of the first equation on their game sheets are filled with a digit.

6. Players find the product of the two numbers on scratch paper.

7. The player with the greater product wins the round.

8. Tiles are returned to the bag and play continues for three more rounds. The player who wins the most rounds wins the game!

Name: ______________________ Date: ______________

# Dunking Decimals

## Game Sheet

**Directions:** Select a number tile. Write the number in one of the five boxes. Continue drawing number tiles until each box has been filled. Multiply the two numbers together. The winner is the player with the larger product.

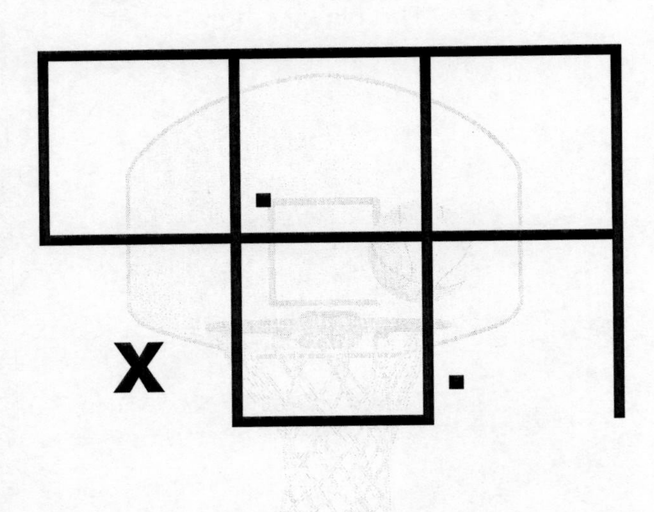

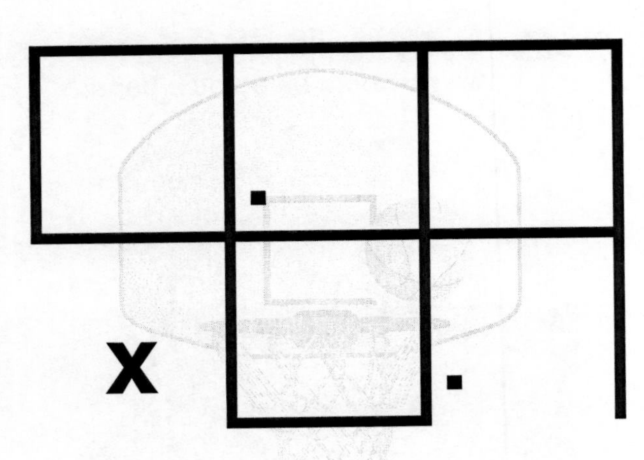

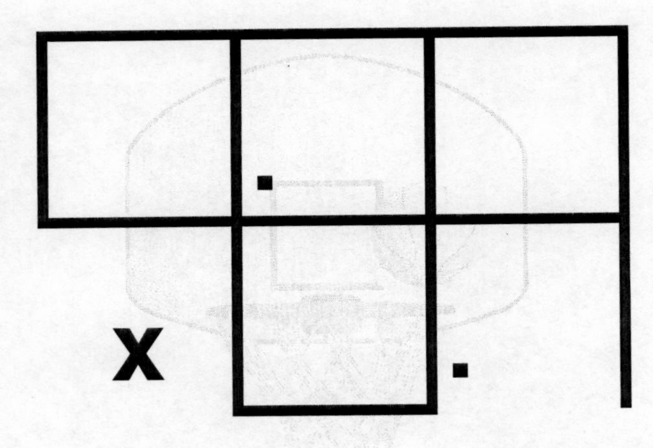

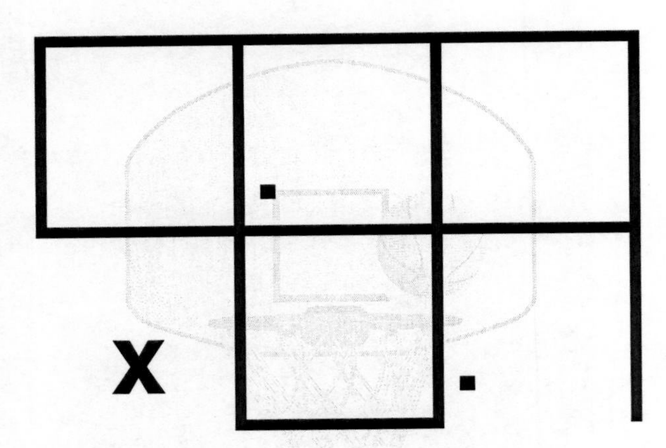

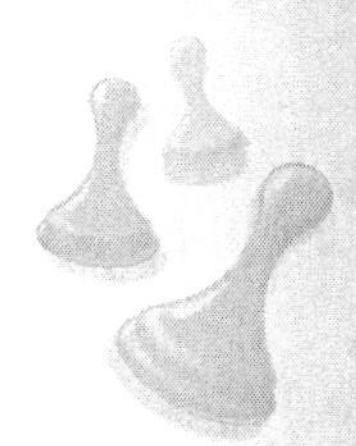

# Dunking Decimals
## Number Tiles

**Directions:** Copy and cut out one set of 0–9 tiles for each pair of players.

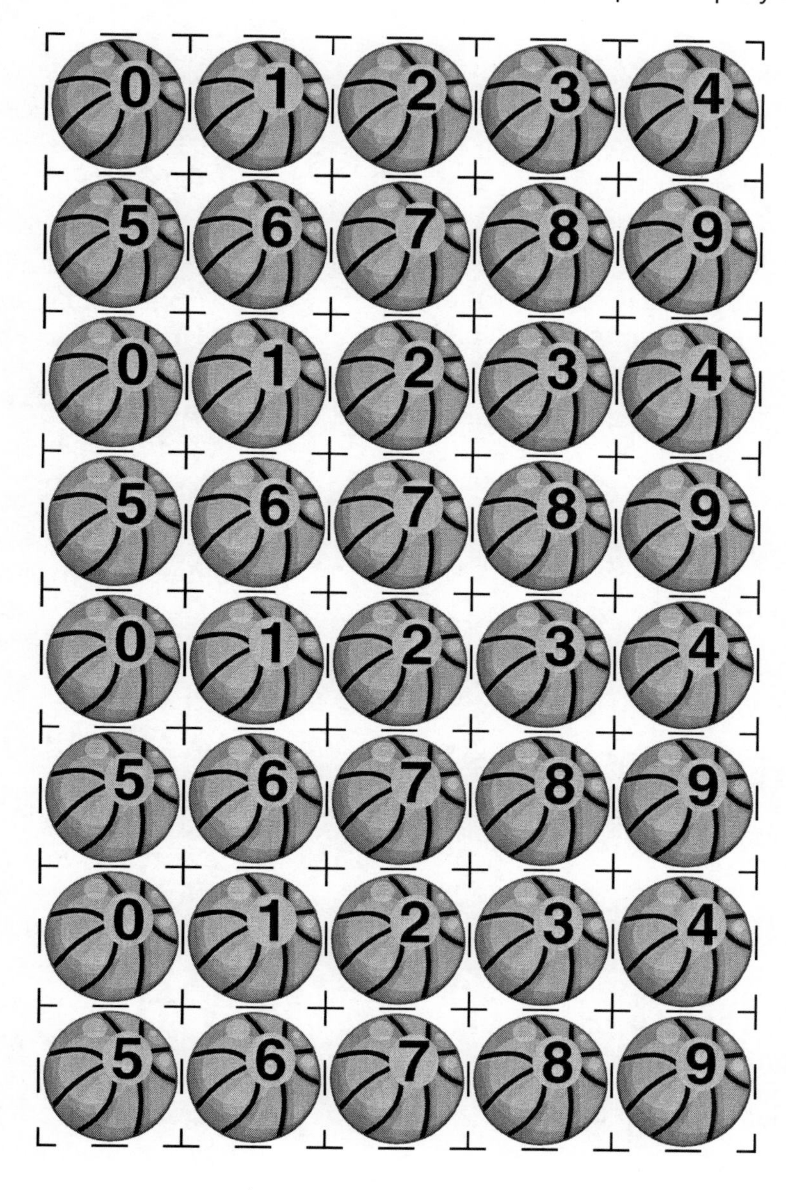

# Mystery Multiples

## Domain

The Number System

## Standard

Find the greatest common factor of two whole numbers less than or equal to 100 and the least common multiple of two whole numbers less than or equal to 12.

## Number of Players

2 to 3 Players

## Materials

- *Mystery Multiples Game Board* (pages 44–45)
- *Mystery Multiples Game Markers* (page 46)
- colored paper (different colors for each player in a group)
- regular decks of cards
- number cubes

## GET PREPARED!

- Copy and cut out one *Mystery Multiples Game Board* for each group of players.
- Copy and cut out the *Mystery Multiples Game Markers* for each group of players, being sure that each player has a different colored paper.
- Collect a deck of cards and a number cube for each group of players.

## Game Directions

1. Distribute materials to players. Each player in a group takes a different colored set of *Mystery Multiples Game Markers*.
2. Players take turns rolling a number cube. The player who rolls the lowest number is Player 1. Players take turns going in clockwise order.

# Mystery Multiples *(cont.)*

3. For this game, the Ace has a value of 1 or 11, the Jack has a value of 12, the Queen has a value of 13, and the King has a value of 14.

4. The deck cards are placed facedown on the play area.

5. To begin, Player 1 draws a card from the deck. The number on the card designates a common factor of two numbers.

6. Player 1 finds two multiples on the *Mystery Multiples Game Board* that have this specific common factor and covers them with their markers. Only two numbers can be covered per turn. The card is placed facedown in a discard pile.

7. Players 2 and 3 repeat steps 5 and 6.

8. A player who cannot cover two multiples loses that turn and places the card in the discard pile.

9. The game ends when all cards are drawn. The winner is the player with the most markers on the game board.

# Mystery Multiples

## Game Board

**Directions:** Copy and cut out game board. Tape it to the game board on page 45.

**Mystery**

| | | |
|---|---|---|
| 1 | 2 | 3 |
| 7 | 8 | 9 |
| 13 | 14 | 15 |
| 19 | 20 | 21 |
| 25 | 26 | 27 |
| 31 | 32 | 33 |
| 37 | 38 | 39 |
| 43 | 44 | 45 |

# Mystery Multiples

## Game Board *(cont.)*

tape here

# Multiples

| | | |
|---|---|---|
| 4 | 5 | 6 |
| 10 | 11 | 12 |
| 16 | 17 | 18 |
| 22 | 23 | 24 |
| 28 | 29 | 30 |
| 34 | 35 | 36 |
| 40 | 41 | 42 |
| 46 | 47 | 48 |

# Mystery Multiples

## Game Markers

**Directions:** Copy and cut out the game markers for each group of players. Be sure to copy these markers on a different colored paper for each player in a group.

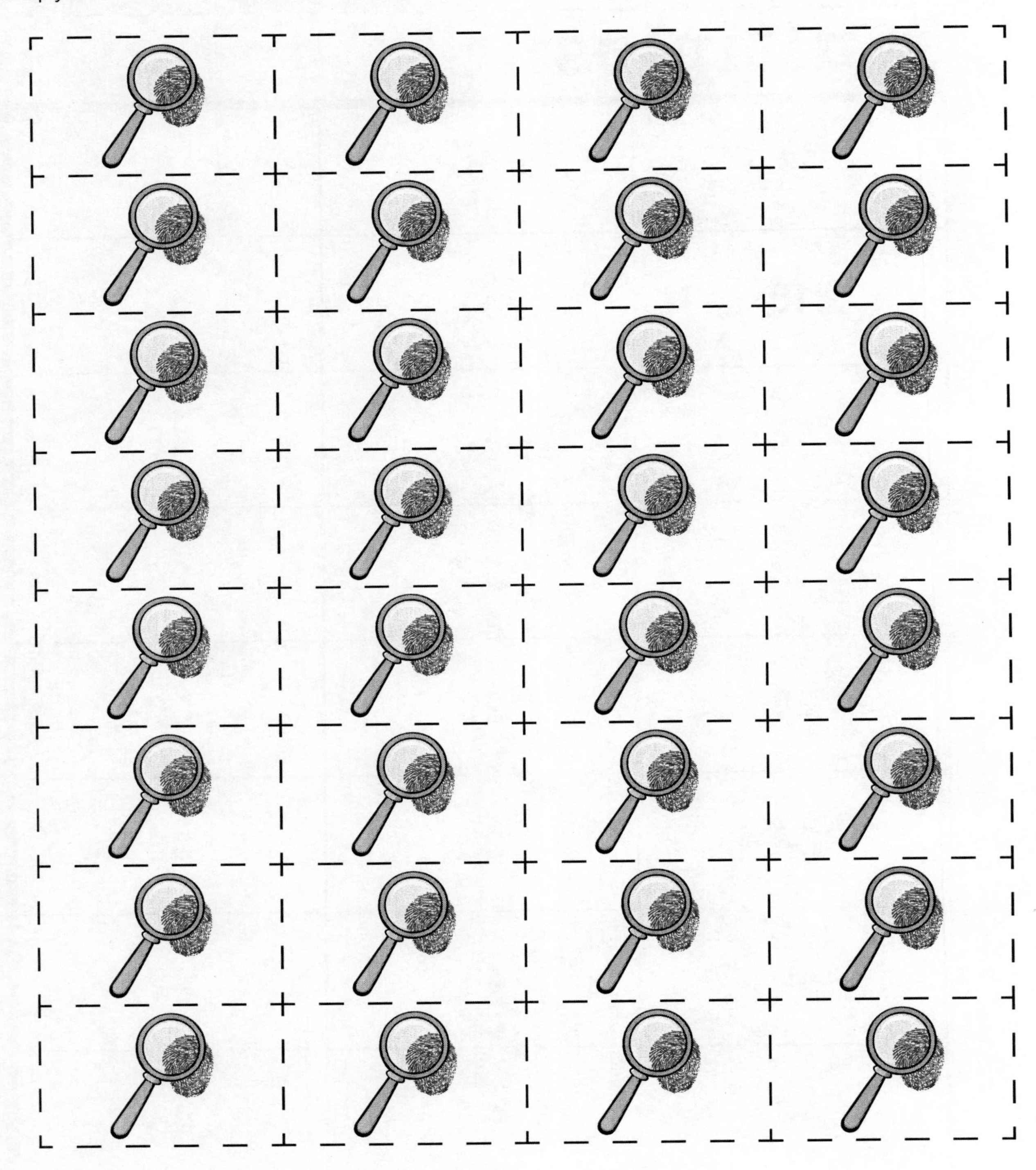

# GCF for the Win

## Domain

The Number System

## Standard

Find the greatest common factor of two whole numbers less than or equal to 100 and the least common multiple of two whole numbers less than or equal to 12.

## Number of Players

2 Players

## Materials

- *GCF for the Win Cards* (pages 49–51)
- *GCF for the Win Game Board* (pages 52–53)
- *GCF for the Win Game Markers* (page 54)
- number cubes

## GET PREPARED!

- Copy and cut out a set of *GCF for the Win Cards* and *GCF for the Win Game Markers* for each pair of players.
- Copy a *GCF for the Win Game Board* for each player.
- Collect a number cube for each pair of players.

## Game Directions

1. Distribute materials to players.
2. Players take turns rolling a number cube. The player who rolls the higher number is Player 1.
3. Players shuffle the cards and place them facedown.
4. To begin, Player 1 draws one of the *GCF for the Win Cards*. The number shown on the card represents the greatest common factor of two numbers on the *GCF for the Win Game Board*.

# GCF for the Win *(cont.)*

5. Player 1 uses his or her *GCF for the Win Game Board* to locate two numbers whose greatest common factor corresponds to the number on the card and covers them with *GCF for the Win Game Markers*. For example, if a player draws a *GCF for the Win Card* showing "50," the player can only cover "50" and "100" on the game board.

6. Player 2 repeats steps 4 and 5.

7. The first player to get four markers in a row, horizontally, vertically, or diagonally is the winner.

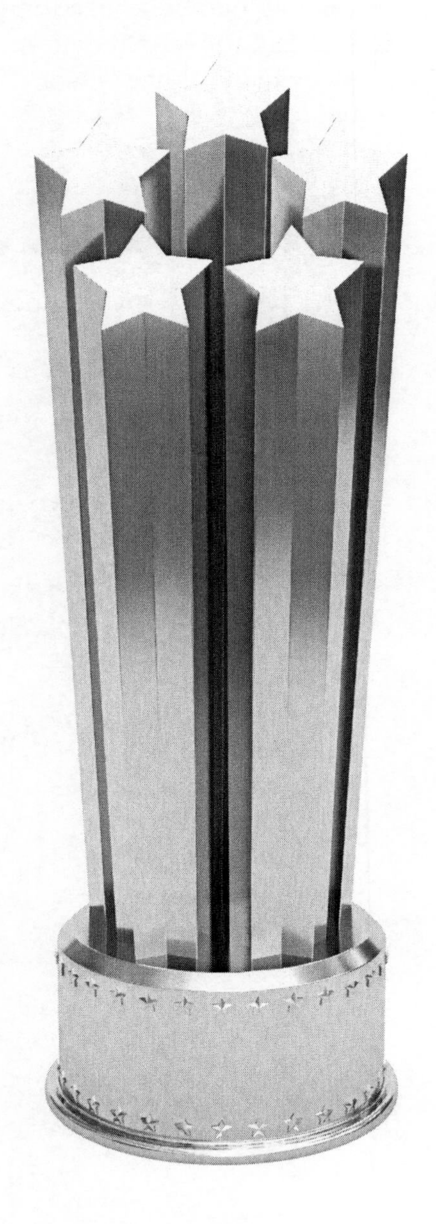

# GCF for the Win

## Cards

**Directions:** Copy and cut out a set of cards for each pair of players.

| | | |
|---|---|---|
| **GCF = 1**  | **GCF = 1**  | **GCF = 1**  |
| **GCF = 1**  | **GCF = 1**  | **GCF = 1**  |
| **GCF = 2**  | **GCF = 2**  | **GCF = 2**  |
| **GCF = 3**  | **GCF = 3**  | **GCF = 3**  |

# GCF for the Win

## Cards *(cont.)*

| | | |
|---|---|---|
| **GCF = 4**  | **GCF = 4**  | **GCF = 4**  |
| **GCF = 5**  | **GCF = 6**  | **GCF = 8**  |
| **GCF = 9**  | **GCF = 10**  | **GCF = 10**  |
| **GCF = 12**  | **GCF = 14**  | **GCF = 15**  |

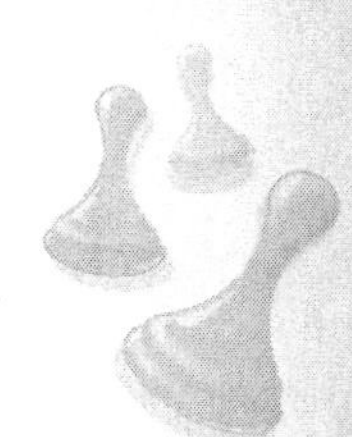

# GCF for the Win

## Cards *(cont.)*

| | | |
|---|---|---|
| GCF = 18  | GCF = 50  | GCF = 11  |
| GCF = 20  | GCF = 2  | GCF = 2  |
| GCF = 5  | GCF = 5  | GCF = 9  |
| GCF = 6  | GCF = 8  | GCF = 7  |
| GCF = 4  | GCF = 3  | GCF = 6  |

# GCF for the Win

## Game Board

**Directions:** Copy and cut out game board. Tape it to the game board on page 53.

GCF for

| | | |
|---|---|---|
| 72 | 49 | 28 |
| 36 | 12 | 26 |
| 15 | 42 | 91 |
| 20 | 99 | 27 |
| 48 | 60 | 64 |
| 77 | 82 | 33 |
| 95 | 46 | 52 |

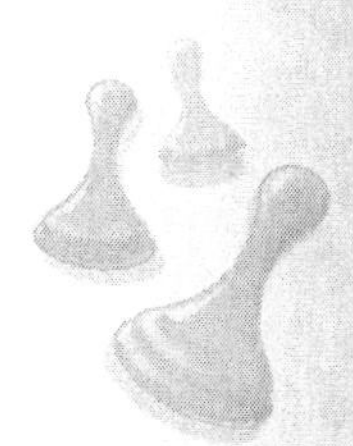

# GCF for the Win

## Game Board *(cont.)*

tape here

## the Win

| | | |
|---|---|---|
| 100 | 18 | 50 |
| 16 | 44 | 35 |
| 88 | 32 | 30 |
| 9 | 11 | 21 |
| 56 | 70 | 55 |
| 14 | 25 | 63 |
| 86 | 74 | 39 |

# GCF for the Win

## Game Markers

**Directions:** Copy and cut out game markers for each player.

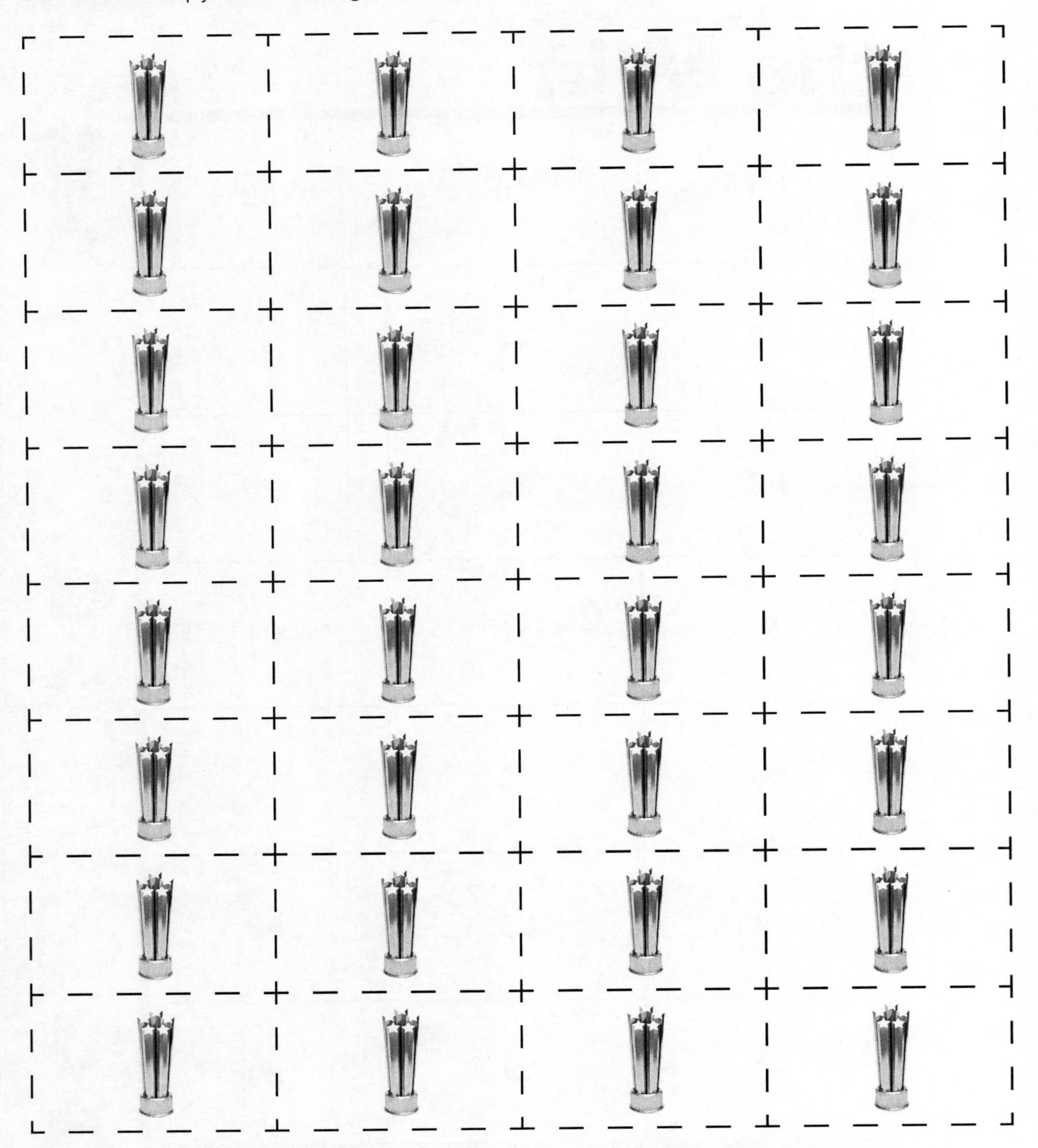

# Quotient Corral

## Domain

The Number System

## Standard

Interpret and compute quotients of fractions, and solve word problems involving division of fractions by fractions.

## GET PREPARED!

- Copy a *Quotient Corral Game Sheet* for each player.
- Collect a number cube for each group of players.

## Number of Players

2 to 3 Players

## Materials

- *Quotient Corral Game Sheet* (page 57)
- number cubes

## Game Directions

1. Distribute materials to players.
2. Players take turns rolling a number cube. The player who rolls the lowest number is Player 1. Players take turns going in clockwise order.
3. To begin, Player 1 tosses the number cube and writes the number in any of the four squares of the first equation on his or her *Quotient Corral Game Sheet*. For example:

$$\frac{\boxed{2}}{\square} \div \frac{\square}{\square} = \underline{\qquad}$$

4. Player 2 tosses the number cube and writes the number in any of the four squares of the first equation on his or her game sheet.

# Quotient Corral *(cont.)*

5. The game continues in the same manner until players have all squares of the first equation filled.

6. Each player calculates the quotient for the division problem.

7. The winner of a round is the player with the greatest quotient.

8. Play continues for four rounds.

**Alternative Directions:**
As an alternative, make the winner the player with the least quotient.

Name:______________________ Date:______________

# Quotient Corral

## Game Sheet

**Directions:** Roll the number cube and fill in the squares of each equation to create fractions.

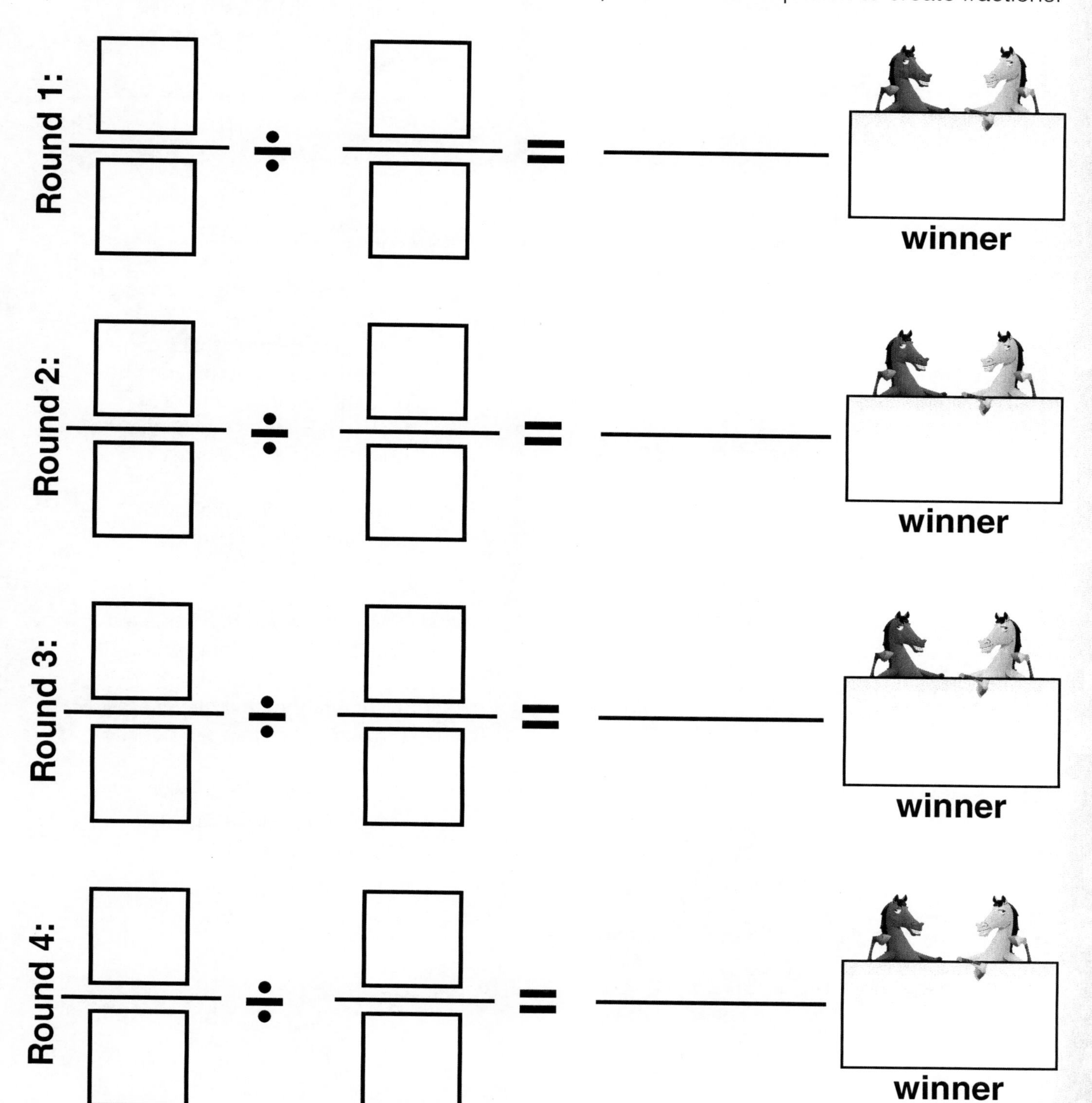

# Zero to Hero

## Domain 

The Number System

## Standard 

Recognize opposite signs of numbers as indicating locations on opposite sides of 0 on the number line; recognize that the opposite of the opposite of a number is the number itself, e.g., $-(-3) = 3$, and that 0 is its own opposite.

## Number of Players 

2 to 3 Players

## Materials 

- regular decks of cards without face cards
- number cubes

## GET PREPARED!

Collect a deck of cards and a number cube for each group of players.

## Game Directions

1. Distribute materials to players.
2. Players take turns rolling a number cube. The player who rolls the highest number is Player 1. Players take turns going in clockwise order.
3. Player 1 shuffles the cards and deals seven cards to each player.
4. The remaining deck is placed facedown on the play area.
5. To begin, the next player draws a card from the deck.
6. Black cards have positive values and red cards have negative values.

 

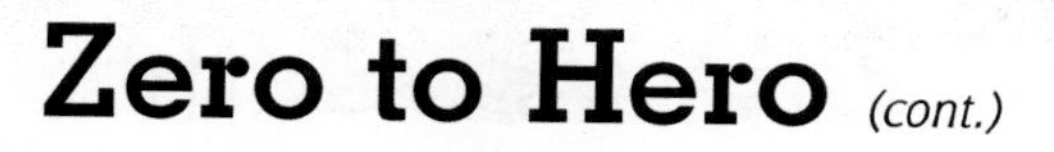

# Zero to Hero *(cont.)*

7. The object of the game is to get rid of all cards by combining two cards so that the sum of the numbers (positive and negative) is zero. For example, a pair containing a black eight and a red eight has a sum of zero and can be laid down on the play area.

8. If a player has opposites, he or she places the two cards face up, and then discards one card.

9. Players may either pick up the top discarded card or a card from the deck.

10. The first player to play all cards is the winner.

# Rational Order

## Domain 

The Number System

## Standard 

Find and position integers and other rational numbers on a horizontal or vertical number line diagram; find and position pairs of integers and other rational numbers on a coordinate plane.

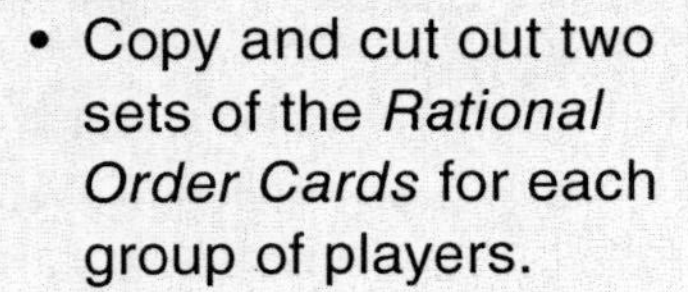


- Copy and cut out two sets of the *Rational Order Cards* for each group of players.
- Copy the *Rational Order Number Line* for each group of players.
- Collect a number cube for each group of players.

## Number of Players 

2 to 3 Players

## Materials 

- *Rational Order Cards* (pages 62–65)
- *Rational Order Number Line* (page 66)
- number cubes

## Game Directions

1. Distribute materials to players.
2. Players take turns rolling a number cube. The player who rolls the lowest number is Player 1. Players take turns going in clockwise order.
3. To begin, Player 1 shuffles the *Rational Order Cards* and deals five cards to each player.
4. The remaining cards are placed facedown and one card is turned over as the starting card and is placed on the *Rational Order Number Line*. If a "1" is drawn, Player 1 draws a different card to place on the number line.

# Rational Order *(cont.)*

5. Play begins by the next player placing a rational number that is greater than or less than the rational number showing on the *Rational Order Number Line*.

6. If the player cannot make a play, he or she discards a card from his or her hand and then draws a *Rational Order Card*. Play passes to the next player.

7. The next player must play a rational number on the *Rational Order Number Line* that is greater than or less than the two rational numbers already shown, not a rational number between the two.

8. The player who plays all five of his or her *Rational Order Cards* on the *Rational Order Number Line,* or has the fewest cards left when all cards have been drawn, is the winner.

# Rational Order

## Cards

**Directions:** Copy and cut out two sets of cards for each group of players.

| | | | |
|---|---|---|---|
| $\frac{1}{10}$ ↓ | $\frac{3}{10}$ ↓ | $\frac{7}{10}$ ↓ | $\frac{9}{10}$ ↓ |
| $\frac{1}{8}$ ↓ | $\frac{3}{8}$ ↓ | $\frac{5}{8}$ ↓ | $\frac{7}{8}$ ↓ |
| $\frac{1}{6}$ ↓ | $\frac{5}{6}$ ↓ | $\frac{1}{5}$ ↓ | $\frac{2}{5}$ ↓ |

 

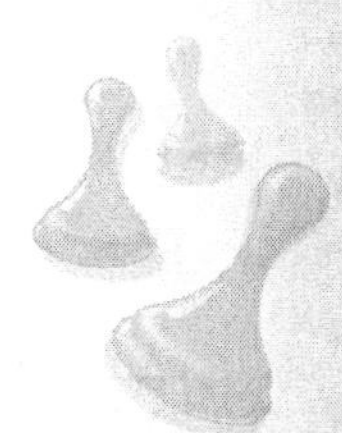

# Rational Order

## Cards *(cont.)*

| | | | |
|---|---|---|---|
| $\frac{3}{5}$ ↓ | $\frac{4}{5}$ ↓ | $\frac{1}{4}$ ↓ | $\frac{3}{4}$ ↓ |
| $\frac{1}{3}$ ↓ | $\frac{2}{3}$ ↓ | $\frac{1}{2}$ ↓ | $\frac{1}{2}$ ↓ |
| $\frac{1}{3}$ ↓ | $\frac{2}{3}$ ↓ | $\frac{1}{4}$ ↓ | $\frac{3}{4}$ ↓ |

# Rational Order

## Cards *(cont.)*

| | | | |
|---|---|---|---|
| $\frac{1}{5}$ ↓ | $\frac{2}{5}$ ↓ | $\frac{3}{5}$ ↓ | $\frac{4}{5}$ ↓ |
| $\frac{1}{6}$ ↓ | $\frac{5}{6}$ ↓ | $\frac{1}{8}$ ↓ | $\frac{3}{8}$ ↓ |
| $\frac{5}{8}$ ↓ | $\frac{7}{8}$ ↓ | $\frac{1}{10}$ ↓ | $\frac{3}{10}$ ↓ |

# Rational Order

## Cards *(cont.)*

| | | | |
|---|---|---|---|
| $\frac{7}{10}$ ↓ | $\frac{9}{10}$ ↓ | $\frac{10}{10}$ ↓ | $\frac{8}{8}$ ↓ |
| $\frac{6}{6}$ ↓ | $\frac{5}{5}$ ↓ | $\frac{4}{4}$ ↓ | $\frac{3}{3}$ ↓ |
| $\frac{2}{2}$ ↓ | 1 ↓ | 1 ↓ | 1 ↓ |

# Rational Order Number Line

**Directions:** Copy and cut out the three number line sections and tape together.

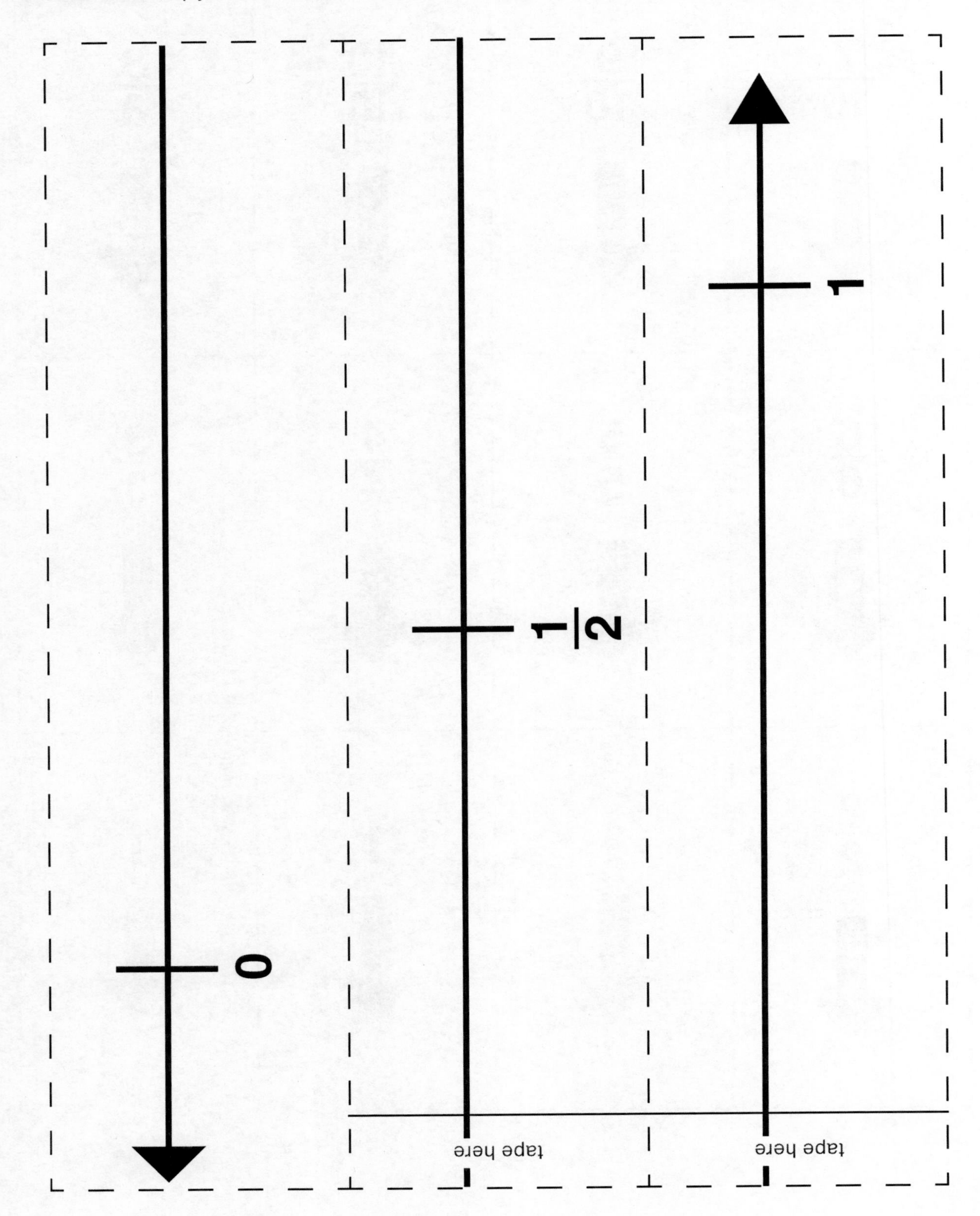

# Express Yourself with Expressions

## Domain

Expressions and Equations

## Standards

Write, read, and evaluate expressions in which letters stand for numbers.

Write expressions that record operations with numbers and with letters standing for numbers.

## Number of Players

2 to 3 Players

## Materials

- *Variable Expression Cards* (pages 69–71)
- *Word Expression Cards* (pages 72–74)
- colored paper (2 different colors)
- number cubes

## GET PREPARED!

- Using one color paper, copy and cut out a set of *Variable Expression Cards* for each group of players. Using another color, copy and cut out a set of *Word Expression Cards* for each group of players.
- Collect a number cube for each group of players.

## Game Directions

1. Distribute materials to players.
2. Players take turns rolling a number cube. The player who rolls the lowest number is Player 1. Players take turns going in a counterclockwise order.
3. Player 1 shuffles the cards and places them facedown on the play area, separating the *Word Expression Cards* from the *Variable Expression Cards*.
4. To begin, Player 1 turns over one card from each pile.

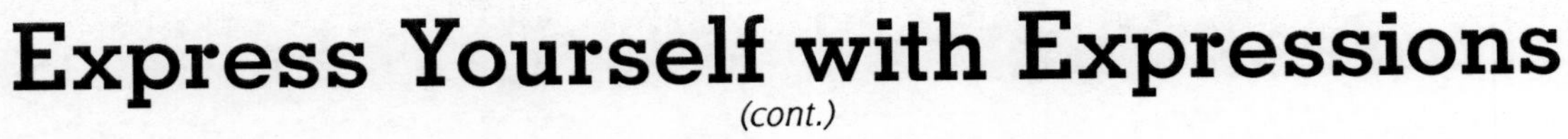

# Express Yourself with Expressions

*(cont.)*

5. If the cards show equivalent expressions, the player takes the two cards from the playing area and places them in his or her winning pile. Otherwise, the cards are placed faceup in the play area.

6. Players 2 and 3 repeat steps 4 and 5.

7. Play continues in the same manner until all cards have been matched.

8. The winner is the player with the most matching cards.

# Variable Expression
## Cards

**Directions:** Copy and cut out the cards for each group of players.

# Variable Expression

## Cards *(cont.)*

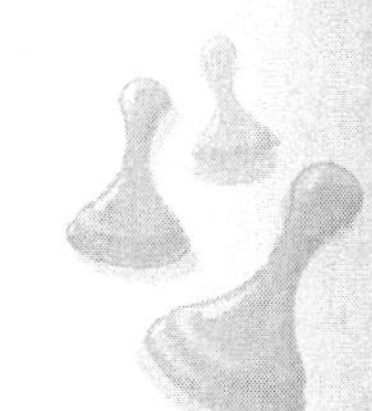

# Variable Expression

## Cards *(cont.)*

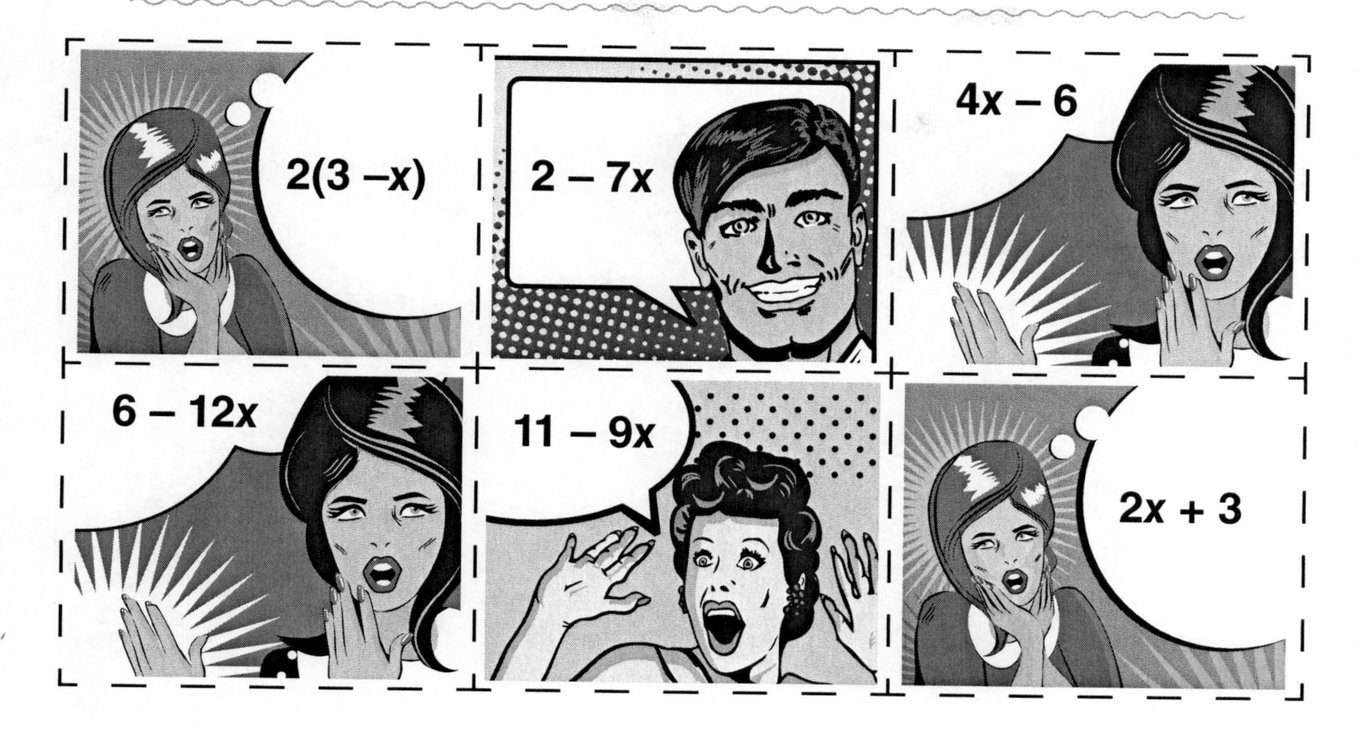

# Word Expression
## Cards

**Directions:** Copy and cut out the cards for each group of players.

# Word Expression
## Cards *(cont.)*

# Word Expression

## Cards *(cont.)*

# Most Valuable 7

## Domain

Expressions and Equations

## Standard

Evaluate expressions at specific values of their variables. Include expressions that arise from formulas used in real-world problems. Perform arithmetic operations, including those involving whole number exponents, in the conventional order when there are no parentheses to specify a particular order (Order of Operations).

## GET PREPARED!

- Copy and cut out a *Most Valuable 7 Game Board* for each player.
- Copy and cut out a set of *Ordered Pair Cards* and the *Most Valuable 7 Game Markers* for each pair of players.
- Collect a number cube for each pair of players.

## Number of Players

2 Players

## Materials

- *Most Valuable 7 Game Board* (pages 77–78)
- *Ordered Pair Cards* (pages 79–80)
- *Most Valuable 7 Game Markers* (page 81)
- number cube

## Game Directions

1. Distribute materials to players.
2. Players take turns rolling a number cube. The player who rolls the higher number is Player 1.
3. To begin, Player 1 shuffles the cards and places them facedown on the play area.
4. Then, Player 1 draws one of the *Ordered Pair Cards*.
5. Player 1 locates the corresponding cell on the *Most Valuable 7 Game Board* and evaluates the expression when $x = 7$. For example, the expression in the cell (3, 4) is $4x + 1$. The value of the expression at $x = 7$ is 29.

# Most Valuable 7 *(cont.)*

6. If correct, Player 1 places a *Most Valuable 7 Game Marker* on the cell.

7. Player 2 repeats steps 4 to 6.

8. Players take turns drawing an *Ordered Pair Card* and evaluating the corresponding expression when $x = 7$. The game is over when a player gets four markers in a row (horizontally, vertically, or diagonally).

9. Players find the sum of all expressions covered on their game board.

10. The player whose sum is greater wins.

**Alternative Directions:**

As an alternative, players evaluate the algebraic expressions at a different value for $x$.

# Most Valuable 7

## Game Board

**Directions:** Copy and cut out game board. Tape it to the game board on page 78.

# Most Valua

| | | | |
|---|---|---|---|
| 6 | 3x – 2 | x – 8 | 5x – 3 |
| 5 | 6x – 3 | 2x + 5 | x + 7 |
| 4 | x – 8 | 5x + 9 | 4x + 1 |
| 3 | 2x – 11 | 4x + 3 | 6x – 7 |
| 2 | 4x + 9 | 2 – 3x | 3x + 8 |
| 1 | 5x + 2 | –6x + 11 | 2x – 9 |
| | 1 | 2 | 3 |

# Most Valuable 7

## Game Board *(cont.)*

ble 7

| | | |
|---|---|---|
| $6x + 5$ | $2x + 7$ | $5 - 4x$ |
| $4x - 5$ | $3x - 3$ | $-5x + 7$ |
| $2x - 7$ | $2 - 3x$ | $-6x + 12$ |
| $5x - 6$ | $x - 12$ | $3x - 13$ |
| $7 - x$ | $6x + 1$ | $2x + 5$ |
| $14 - 3x$ | $4x - 9$ | $x + 8$ |
| 4 | 5 | 6 |

tape here

# Ordered Pair

## Cards

**Directions:** Copy and cut out the cards for each pair of players.

| | | | | | |
|---|---|---|---|---|---|
| (1, 1) | (1, 2) | (1, 3) | (1, 4) | (1, 5) | (6, 6) |
| (1, 6) | (2, 1) | (2, 2) | (2, 3) | (2, 4) | (6, 3) |
| (2, 5) | (2, 6) | (3, 1) | (3, 2) | (3, 3) | (6, 4) |
| (3, 4) | (3, 5) | (3, 6) | (4, 6) | (5, 6) | (6, 5) |

# Ordered Pair

## Cards *(cont.)*

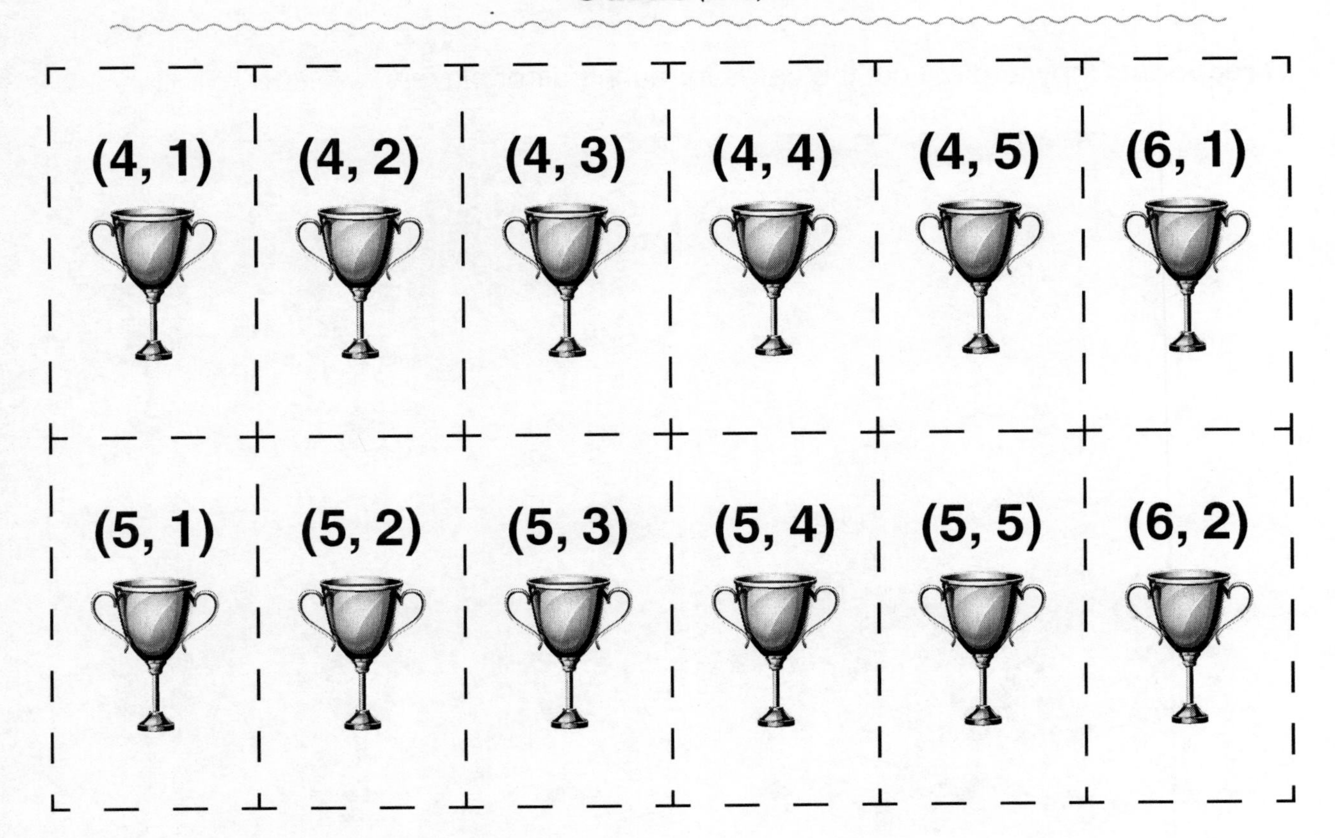

# Most Valuable 7

## Game Markers

**Directions:** Copy and cut out the game markers for each group of players.

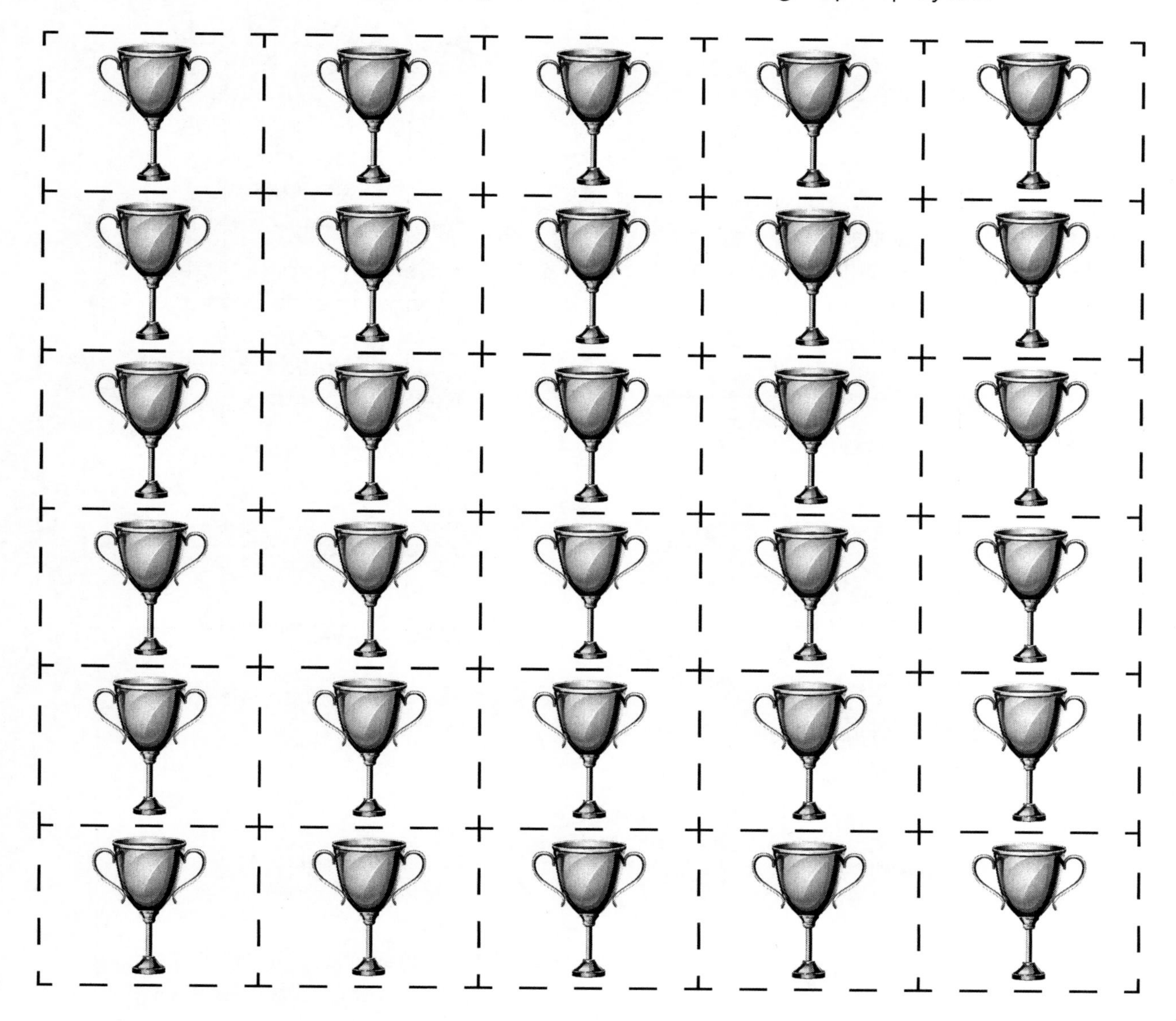

# Distribution Concentration

## Domain

Expressions and Equations

## Standard

Apply the properties of operations to generate equivalent expressions.

## Number of Players

2 to 3 Players

## Materials

- *Factored Concentration Cards* (pages 84–85)
- *Distributed Concentration Cards* (pages 86–87)
- colored paper (two different colors)
- number cubes

## GET PREPARED!

- Copy and cut out a set of *Factored Concentration Cards* for each group of players on one colored paper.
- Use a different colored paper to copy and cut out a set of *Distributed Concentration Cards* for each group of players.
- Collect a number cube for each group of players.

## Game Directions

1. Distribute materials to players.
2. Players take turns rolling a number cube. The player who rolls the lowest number is Player 1. Players take turns going in a clockwise order.
3. Player 1 shuffles the cards and places them facedown on the play area, separating the factored form from the distributed form.
4. Player 1 draws and turns over one card from the factored form pile and one card from the distributed form pile.

# Distribution Concentration *(cont.)*

5. If the cards show equivalent expressions, the player takes the two cards from the playing area. Otherwise, the cards are placed at the bottom of the pile.
6. Play continues in the same manner until all cards have been matched.
7. The winner is the player with the most cards.

# Factored Concentration
## Cards

**Directions:** Copy and cut out the cards for each group of players.

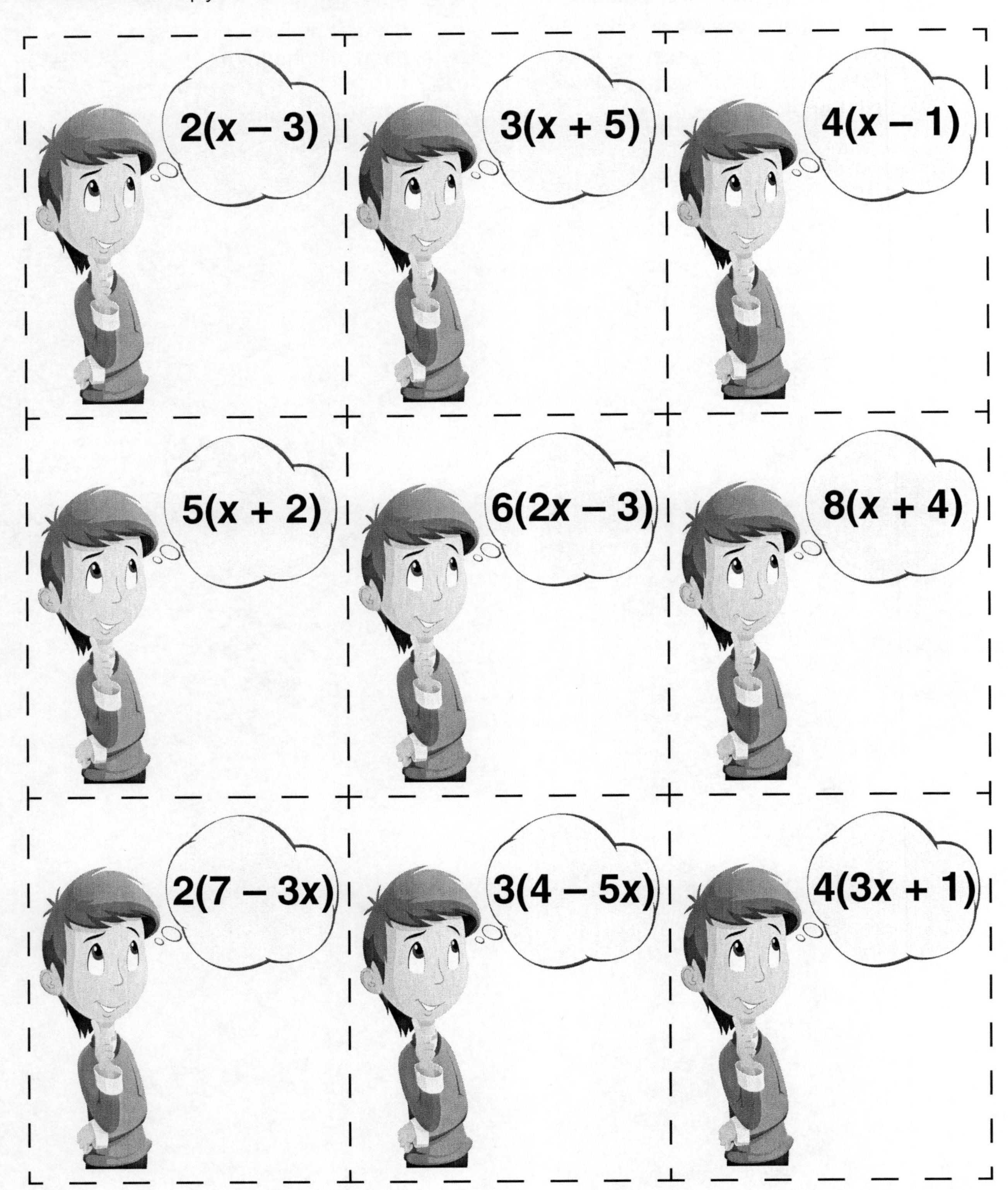


 

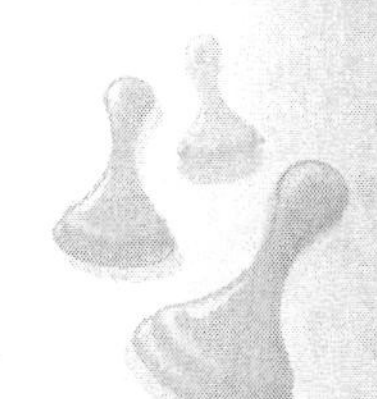

# Factored Concentration

## Cards *(cont.)*

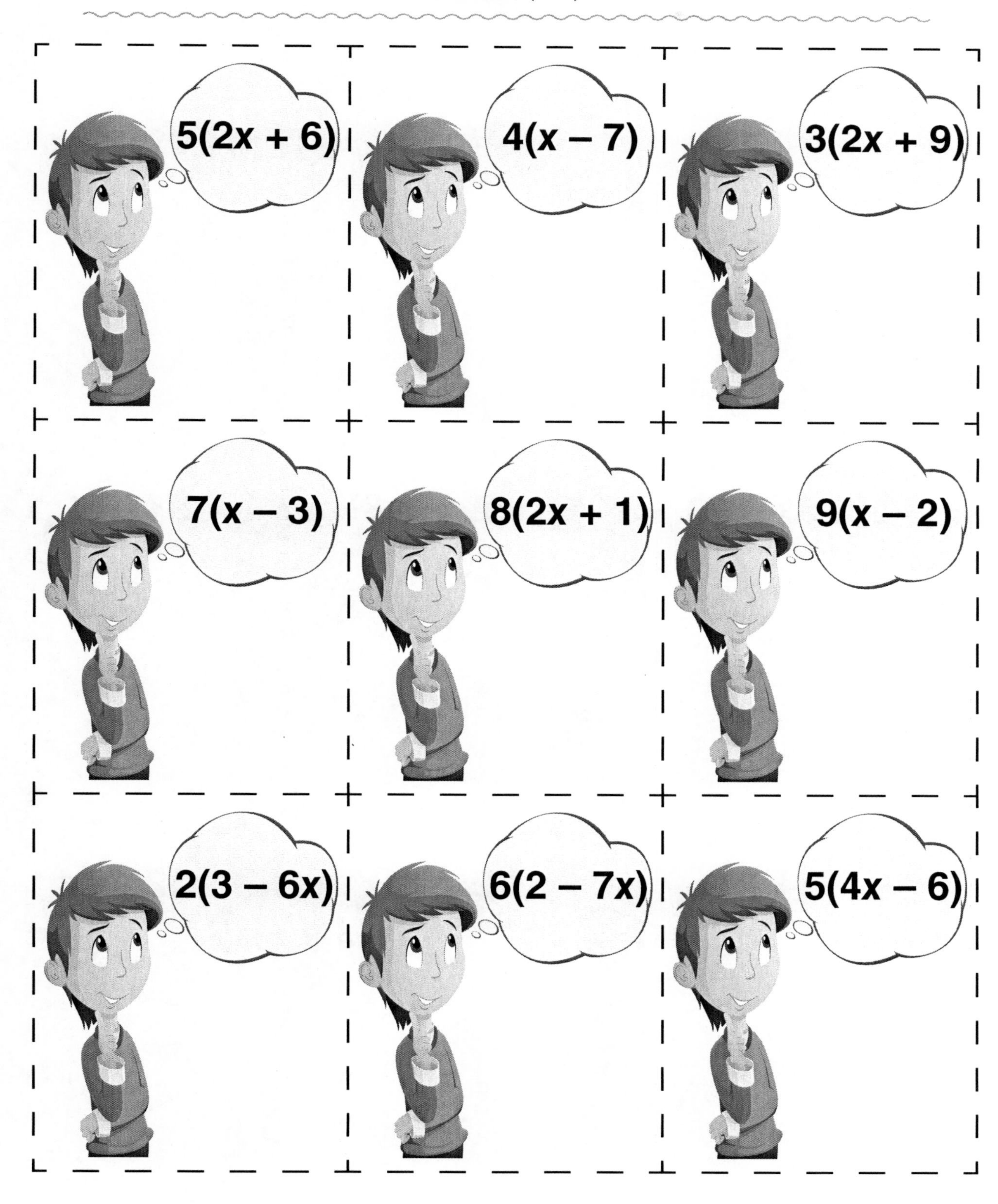

# Distributed Concentration

## Cards

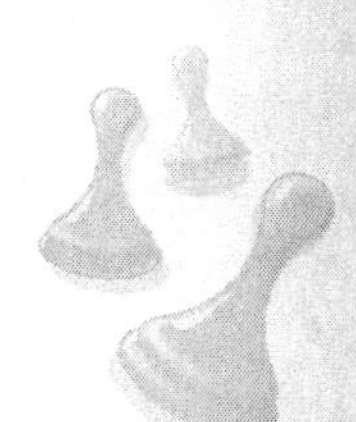

# Distributed Concentration

## Cards *(cont.)*

# Equation Bingo

## Domain

Expressions and Equations

## Standard

Understand solving an equation or inequality as a process of answering a question: which values from a specified set, if any, make the equation or inequality true? Use substitution to determine whether a given number in a specified set makes an equation or inequality true.

## GET PREPARED!

- Copy a set of the *Equation Bingo Game Boards* for each group of players.
- Copy and cut out a set of the *Equation Cards* and *Equation Bingo Markers* for each group of players.
- Collect a number cube for each group of players.

## Number of Players

3 to 4 Players

## Materials

- *Equation Cards Set 1* (page 90)
- *Equation Cards Set 2* (page 91)
- *Equation Bingo Game Board Set 1* (pages 92–95)
- *Equation Bingo Game Board Set 2* (pages 96–99)
- *Equation Bingo Markers* (page 100)
- number cubes

## Game Directions

1. Distribute materials to players.
2. Players take turns rolling a number cube. The player who rolls the lowest number is Player 1. Players take turns going in a counterclockwise order.
3. Player 1 shuffles the *Equation Cards* and places them facedown on the play area.
4. Player 1 draws an *Equation Card* and places it face up for all players to see.

# Equation Bingo *(cont.)*

5. Each player solves the equation and marks the solution on his or her *Equation Bingo Game Board*.

6. Play continues in the same manner, with the next player drawing a card and placing it face up.

7. The first player to get a Tic-Tac-Toe (three markers in a row, horizontally, vertically, or diagonally) is the winner.

8. Players reshuffle the deck and play again, or a different set of *Equation Bingo Game Boards* and *Equation Cards* can be distributed.

# Equation Cards

## Set 1

**Directions:** Copy and cut out the cards for each group of players.

$5x + 3 = 8$

$2x - 5 = 7$

$3x - 2 = 10$

$4x + 5 = 13$

$6x - 8 = 10$

$2x - 12 = 6$

$4x + 7 = 35$

$3x + 4 = 28$

$7x - 20 = 15$

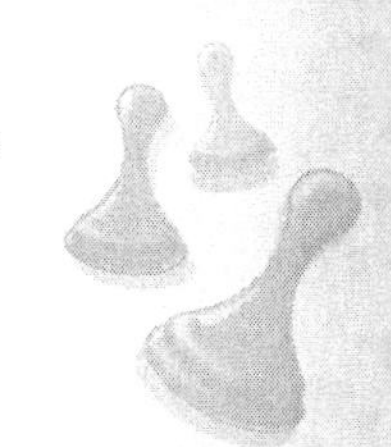

# Equation Cards
## Set 2

**Directions:** Copy and cut out cards for each group of players.

**$7x = 56$**

**$4x = 4$**

**$5x = 30$**

**$2x = 18$**

**$3x = 15$**

**$9x = 27$**

**$8x = 16$**

**$12x = 48$**

**$6x = 42$**

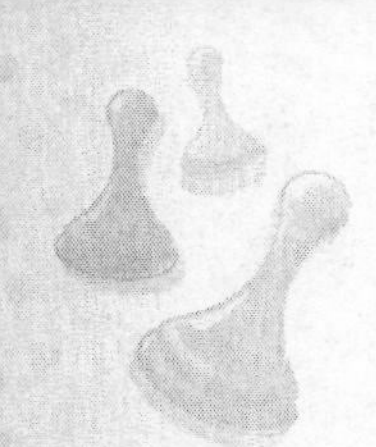

# Equation Bingo

## Game Board Set 1

**Directions:** Draw an equation card. Solve the equation and mark the solution on the game sheet. The winner is the first player to get three in a row horizontally, vertically, or diagonally.

| Equation Bingo Board 1 | | |
|---|---|---|
| 1 | 2 | 3 |
| 4 | 5 | 6 |
| 7 | 8 | 9 |

 

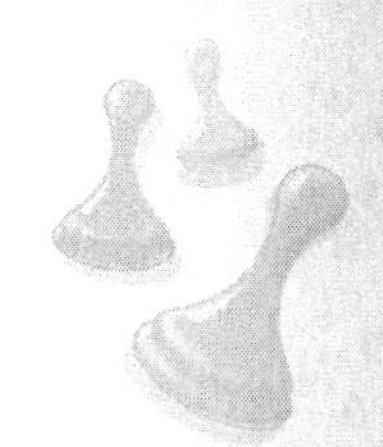

# Equation Bingo

## Game Board Set 1

**Directions:** Draw an equation card. Solve the equation and mark the solution on the game sheet. The winner is the first player to get three in a row horizontally, vertically, or diagonally.

| Equation Bingo<br>Board 2 | | |
|---|---|---|
| **1** | **3** | **5** |
| **7** | **9** | **2** |
| **4** | **6** | **8** |

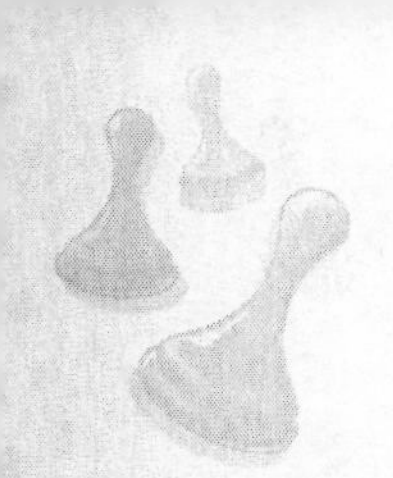

# Equation Bingo

## Game Board Set 1

**Directions:** Draw an equation card. Solve the equation and mark the solution on the game sheet. The winner is the first player to get three in a row horizontally, vertically, or diagonally.

| Equation Bingo<br>Board 3 | | |
|---|---|---|
| 2 | 5 | 8 |
| 1 | 4 | 7 |
| 3 | 6 | 9 |

 

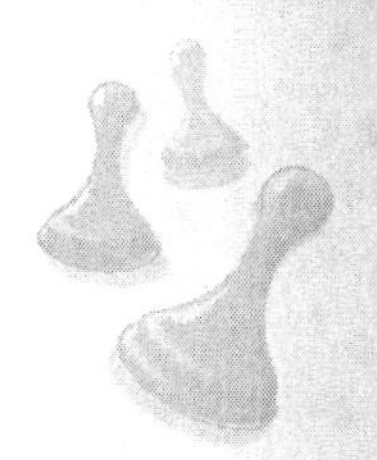

# Equation Bingo

## Game Board Set 1

**Directions:** Draw an equation card. Solve the equation and mark the solution on the game sheet. The winner is the first player to get three in a row horizontally, vertically, or diagonally.

| Equation Bingo<br>Board 4 | | |
|---|---|---|
| **5** | **1** | **9** |
| **6** | **2** | **7** |
| **3** | **8** | **4** |

# Equation Bingo

## Game Board Set 2

**Directions:** Draw an equation card. Solve the equation and mark the solution on the game sheet. The winner is the first player to get three in a row horizontally, vertically, or diagonally.

| Equation Bingo<br>Board 5 | | |
|---|---|---|
| 1 | 2 | 3 |
| 4 | 5 | 6 |
| 7 | 8 | 9 |

 

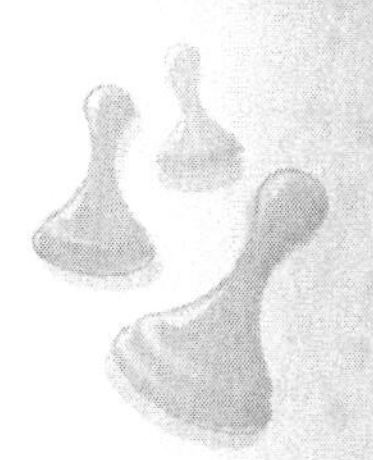

# Equation Bingo

## Game Board Set 2

**Directions:** Draw an equation card. Solve the equation and mark the solution on the game sheet. The winner is the first player to get three in a row horizontally, vertically, or diagonally.

| Equation Bingo<br>Board 6 | | |
|---|---|---|
| **1** | **3** | **5** |
| **7** | **9** | **2** |
| **4** | **6** | **8** |

# Equation Bingo
## Game Board Set 2

**Directions:** Draw an equation card. Solve the equation and mark the solution on the game sheet. The winner is the first player to get three in a row horizontally, vertically, or diagonally.

| Equation Bingo<br>Board 7 | | |
|---|---|---|
| 2 | 5 | 8 |
| 1 | 4 | 7 |
| 3 | 6 | 9 |

 

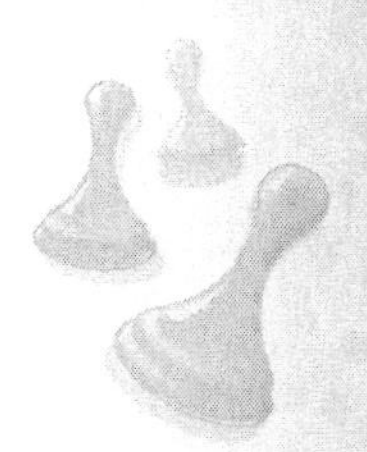

# Equation Bingo

## Game Board Set 2

**Directions:** Draw an equation card. Solve the equation and mark the solution on the game sheet. The winner is the first player to get three in a row horizontally, vertically, or diagonally.

| Equation Bingo Board 8 | | |
|---|---|---|
| 5 | 1 | 9 |
| 6 | 2 | 7 |
| 3 | 8 | 4 |

# Equation Bingo

## Markers

**Directions:** Copy and cut out the game markers for each group of players.

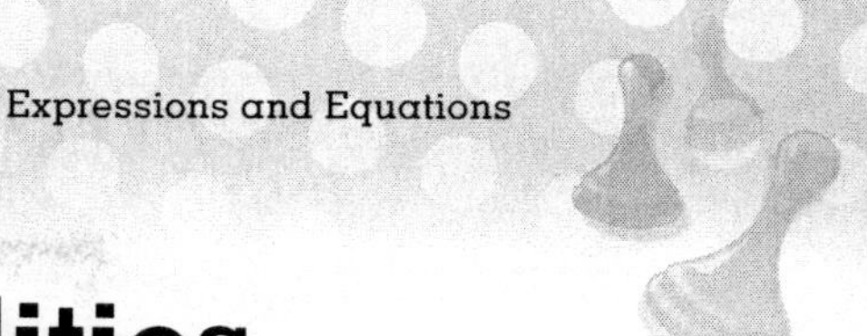

# True or False: Inequalities

## Domain

Expressions and Equations

## Standard

Understand solving an equation or inequality as a process of answering a question: which values from a specified set, if any, make the equation or inequality true? Use substitution to determine whether a given number in a specified set makes an equation or inequality true.

## Number of Players

2 to 3 Players

## Materials

- *True or False: Inequalities Cards* (pages 103–105)
- regular decks of cards
- number cubes

## GET PREPARED!

- Copy and cut out a set of *True or False: Inequalities Cards* for each group of players.
- Collect a deck of cards and a number cube for each group of players.

## Game Directions

1. Distribute materials to players.
2. Players take turns rolling a number cube. The player who rolls the lowest number is Player 1. Players take turns going in a clockwise order.
3. The *True or False: Inequalities Cards* and the regular deck of cards are placed facedown on the play area.
4. To begin, Player 1 turns over a card from each deck. An ace has a value of 1, a Jack has a value of 11, a Queen has a value of 12, and a King has a value of 13.

# True or False: Inequalities *(cont.)*

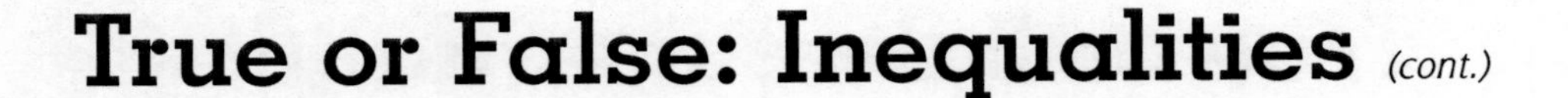

5. If the inequality is true, Player 1 collects the two cards, places them in his or her winning pile, and draws another card from each deck. For example, if Player 1 draws a "6" and "$x < 8$," he or she collects both cards since the inequality is true. If Player 1 draws "5" and "$x < 2$" the inequality is false and play passes to the next player.
6. Player 1 continues turning over a number card and a *True or False: Inequalities Card* until the number does not satisfy the inequality.
7. The *True or False: Inequalities Card* is placed in a discard pile, but the number card remains for the next player.
8. Players 2 and 3 repeat steps 4 to 7.
9. The game ends when all cards are used in either deck.
10. The winner is the player with the most pairs of cards.

# True or False: Inequalities

## Cards

**Directions:** Copy and cut out the cards for each group of players.

| | | |
|---|---|---|
| $x < 1$ | $x < 1$ | $x < 2$ |
| $x < 2$ | $x < 3$ | $x < 3$ |
| $x < 4$ | $x < 4$ | $x < 5$ |
| $x < 5$ | $x < 6$ | $x < 6$ |

# True or False: Inequalities

## Cards *(cont.)*

| | | |
|---|---|---|
| $x < 7$ | $x < 7$ | $x < 8$ |
| $x < 9$ | $x < 10$ | $x < 11$ |
| $x > 1$ | $x > 2$ | $x > 2$ |
| $x > 3$ | $x > 3$ | $x > 4$ |

 

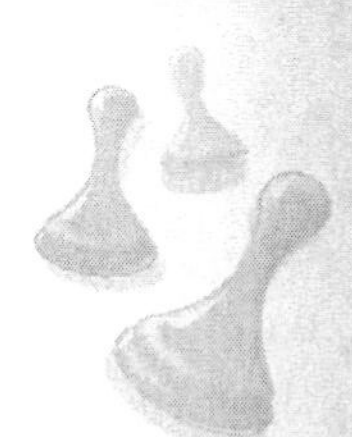

# True or False: Inequalities

## Cards *(cont.)*

| | | |
|---|---|---|
| ? ? $x > 4$ ? ? | ? ? $x > 5$ ? ? | ? ? $x > 5$ ? ? |
| ? ? $x > 6$ ? ? | ? ? $x > 6$ ? ? | ? ? $x > 7$ ? ? |
| ? ? $x > 7$ ? ? | ? ? $x > 8$ ? ? | ? ? $x > 8$ ? ? |
| ? ? $x > 9$ ? ? | ? ? $x > 9$ ? ? | ? ? $x > 10$ ? ? |
| ? ? $x > 10$ ? ? | ? ? $x > 11$ ? ? | ? ? $x > 11$ ? ? |

# High Velocity Volume

## Domain

Geometry

## Standard

Find the volume of a right rectangular prism with fractional edge lengths by packing it with unit cubes of the appropriate unit fraction edge lengths, and show that the volume is the same as would be found by multiplying the edge lengths of the prism. Apply the formulas $V = l\ w\ h$ and $V = b\ h$ to find volumes of right rectangular prisms with fractional edge lengths in the context of solving real-world and mathematical problems. Represent three-dimensional figures using nets made up of rectangles and triangles, and use the nets to find the surface area of these figures.

## Number of Players

2 Players

## Materials

- *High Velocity Volume Game Board* (pages 108–109)
- *High Velocity Volume Game Markers* (page 110)
- *High Velocity Volume Answer Key* (page 133)
- colored paper (two different colors)
- number cubes
- scratch paper

## GET PREPARED

- Copy and cut out a *High Velocity Volume Game Board* for each pair of players.
- Using two different colored papers, copy and cut out the *High Velocity Volume Game Markers*.
- Collect a number cube and scratch paper for each pair of players.

## Game Directions

1. Distribute materials to players.
2. Players take turns rolling a number cube. The player who rolls the lower number is Player 1.
3. Both players place their markers on the "Start" space.
4. Player 1 tosses the number cube and moves forward the indicated number of spaces on the *High Velocity Volume Game Board*.

 

# High Velocity Volume *(cont.)*

5. Player 1 follows the directions indicated in the space by either finding the surface area or volume of the rectangular prism with the measurements provided using scratch paper. If incorrect, players move their *High Velocity Volume Game Markers* back.

6. Player 2 repeats steps 4 and 5.

7. Play continues in the same manner for each player. To win, a player must land exactly on the "Start/Finish" space.

8. Answers can be checked using the *High Velocity Volume Answer Key.*

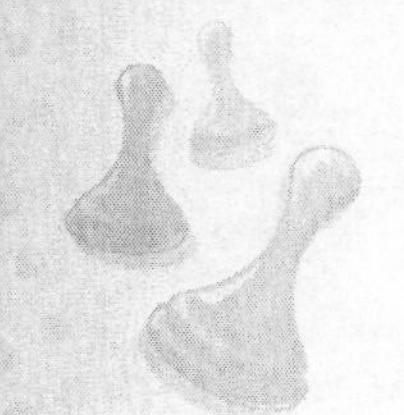



# High Velocity Volume

## Game Board

**Directions:** Copy and cut out the game board. Tape it to the board on page 109.

| | | |
|---|---|---|
| **Surface Area**<br>**L = 5 m**<br>**W = 4 m**<br>**H = 2 m** | **Volume**<br>**L = 4 cm**<br>**W = 2 cm**<br>**H = 2 cm** | **Go Back One** |
| **Volume**<br>**L = 3 ft.**<br>**W = 4 ft.**<br>**H = 2 ft.** | HIGH VELOC | |
| **Surface Area**<br>**L = 2 m**<br>**W = 2 m**<br>**H = 1 m** | | |
| **Volume**<br>**L = 2 m**<br>**W = 2 m**<br>**H = 1 m** | | |
| **↑**<br>**Start**<br>**Finish** | **Surface Area**<br>**L = 4 in.**<br>**W = 4 in.**<br>**H = 4 in.** | **Volume**<br>**L = 5 ft.**<br>**W = 3 ft.**<br>**H = 2 ft.** |

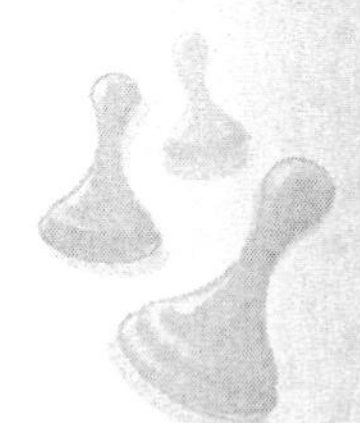

# High Velocity Volume

## Game Board *(cont.)*

| Surface Area<br>L = 4 cm<br>W = 3 cm<br>H = 3 cm | Volume<br>L = 3 in.<br>W = 3 in.<br>H = 3 in. | Surface Area<br>L = 5 in.<br>W = 2 in.<br>H = 3 in. |
|---|---|---|

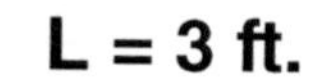

|  |
|---|
| Volume<br>L = 2 ft.<br>W = 3 ft.<br>H = 1 ft. |
| Surface Area<br>L = 3 ft.<br>W = 3 ft.<br>H = 3 ft. |
| Go Back Two |

| Surface Area<br>L = 3 ft.<br>W = 4 ft.<br>H = 2 ft. | Move Ahead Two | Surface Area<br>L = 4 m<br>W = 2 m<br>H = 2 m |
|---|---|---|

tape here

# High Velocity Volume

## Game Markers

**Directions:** Copy and cut out the game markers for each pair of players. Be sure to copy these markers on colored paper, a different color for each player.

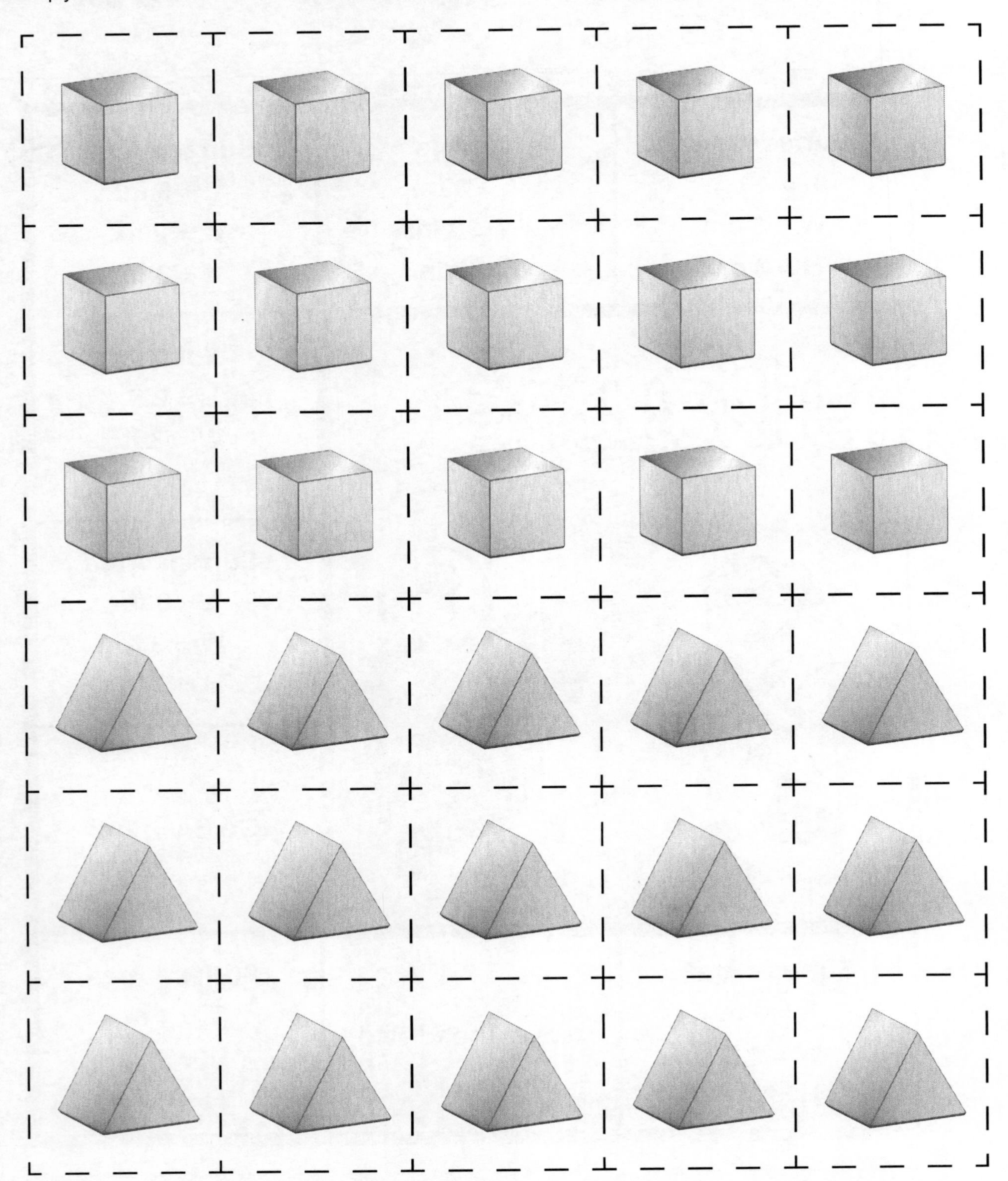

# Triangles

## Domain

Geometry

## Standard

Draw polygons in the coordinate plane given coordinates for the vertices; use coordinates to find the length of a side joining points with the same first coordinate or the same second coordinate. Apply these techniques in the context of solving real-world and mathematical problems.

## Number of Players

2–3 Players

## Materials

- *Triangles Grid Sheet* (page 113)
- number cubes

## GET PREPARED!

- Copy a *Triangles Grid Sheet* for each player.
- Collect a number cube for each group of players.

## Game Directions

1. Distribute materials to players.
2. Players take turns rolling a number cube. The player who rolls the highest number is Player 1. Players take turns going in a clockwise order.
3. To begin, Player 1 tosses the number cube. The number represents the area of a triangle in square units. For example, if the number cube lands on 4, a four-square area should be represented on the *Triangles Grid Sheet*.
4. Player 1 draws a triangle on the grid that has an area corresponding to the number on the cube.

# Triangles *(cont.)*

5. The other players repeat steps 3 and 4.

6. A player that cannot draw a triangle on the grid sheet without overlapping with another triangle is out of the game.

7. Play continues until only one player remains. That player is the winner.

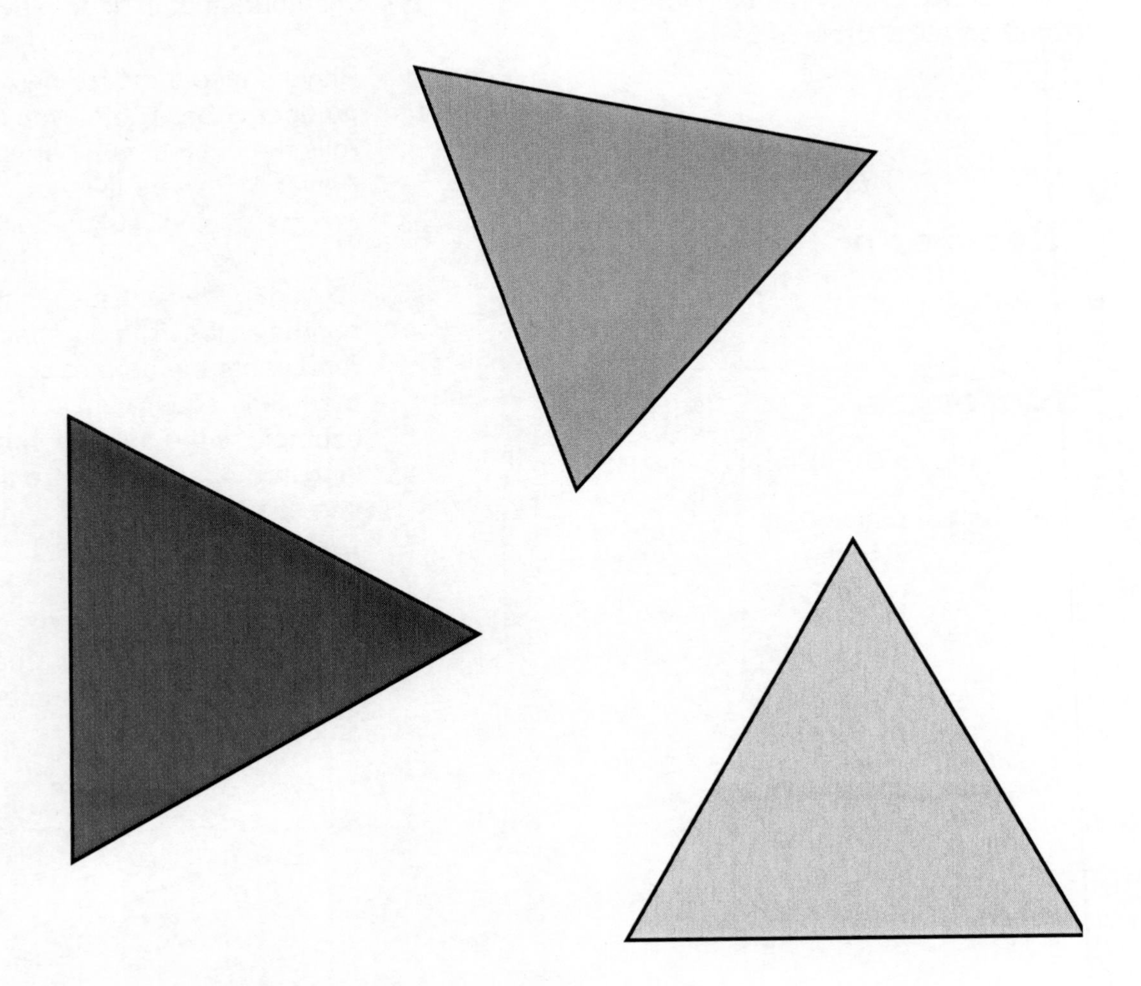

Name:________________________________ Date:________________

# Triangles

## Grid Sheet

**Directions:** Roll the number cube. Draw a triangle that has the area of the number that was rolled on the number cube.

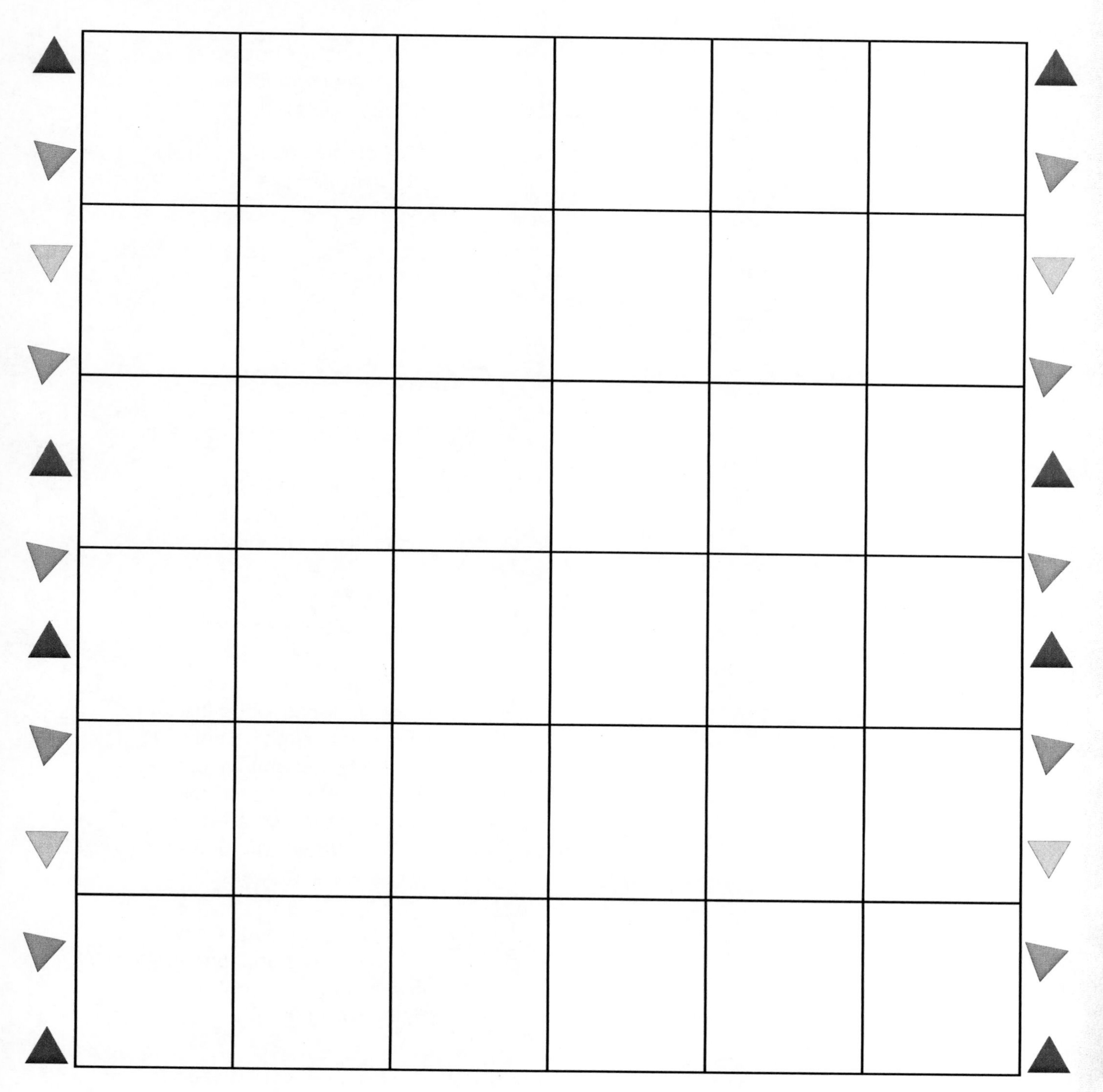

# Mean Wins

## Domain 

Statistics and Probability

## Standard 

Display numerical data in plots on a number line, including dot plots, histograms, and box plots.

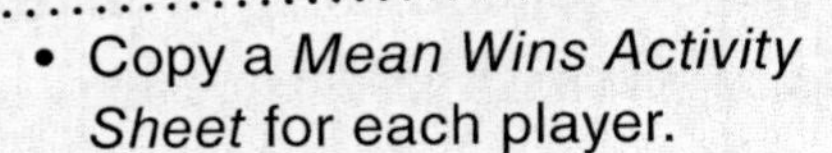

## GET PREPARED!

- Copy a *Mean Wins Activity Sheet* for each player.
- Copy and cut out the *Mean Wins Markers* for each group of players.
- Collect two number cubes and scratch paper for each group of players.

## Number of Players 

2 to 3 Players

## Materials 

- *Mean Wins Activity Sheet* (page 116)
- *Mean Wins Markers* (page 117)
- number cubes
- scratch paper

## Game Directions

1. Distribute materials to players.
2. Players take turns rolling a number cube. The player who rolls the highest number is Player 1. Players take turns going in a counterclockwise order.
3. To begin, Player 1 tosses the number cubes, finds the sum of the two numbers showing on the faces, and covers one square in the appropriate column of the *Mean Wins Activity Sheet* with one of the *Mean Wins Markers*.
4. The game continues in the same manner until one player completely fills a column for any particular sum.

# Mean Wins *(cont.)*

5. At that point, each player uses scratch paper to determine the mean of all the sums covered. For example,

| | | | | | | | | | | |
|---|---|---|---|---|---|---|---|---|---|---|
| | | | | | x | | | | | |
| | | | | | x | x | | | | |
| | | x | | x | x | x | | | x | |
| x | x | x | | x | x | x | x | | x | x |
| 2 | 3 | 4 | 5 | 6 | 7 | 8 | 9 | 10 | 11 | 12 |

Sum = $(1 \times 2) + (1 \times 3) + (2 \times 4) + (2 \times 6) + (4 \times 7) + (3 \times 8) + (1 \times 9) + (2 \times 11) + (1 \times 12) = 120$

Mean = $120 \div 17 = 7.1$

6. The player with the greatest mean is the winner of the round.

7. Players play three rounds to determine an overall winner.

8. If three players are playing and each one wins a round, an additional round is played to determine the winner.

Name: ____________________________ Date: ____________________

# Mean Wins

## Activity Sheet

**Directions:** Roll the number cubes. Find the sum of the number cubes. Cover the sum on the activity sheet. Continue play until one player has filled up a number column. Find the mean of the numbers covered on the activity sheet. The player with the greatest mean is the winner.

| | | | | | | | | | | |
|---|---|---|---|---|---|---|---|---|---|---|
| | | | | | | | | | | |
| | | | | | | | | | | |
| | | | | | | | | | | |
| | | | | | | | | | | |
| 2 | 3 | 4 | 5 | 6 | 7 | 8 | 9 | 10 | 11 | 12 |

 

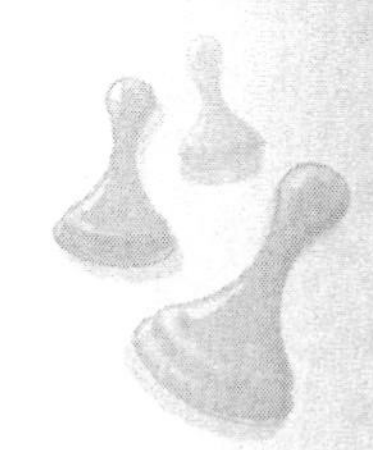

# Mean Wins
## Markers

**Directions:** Copy and cut out the game markers for each group of players.

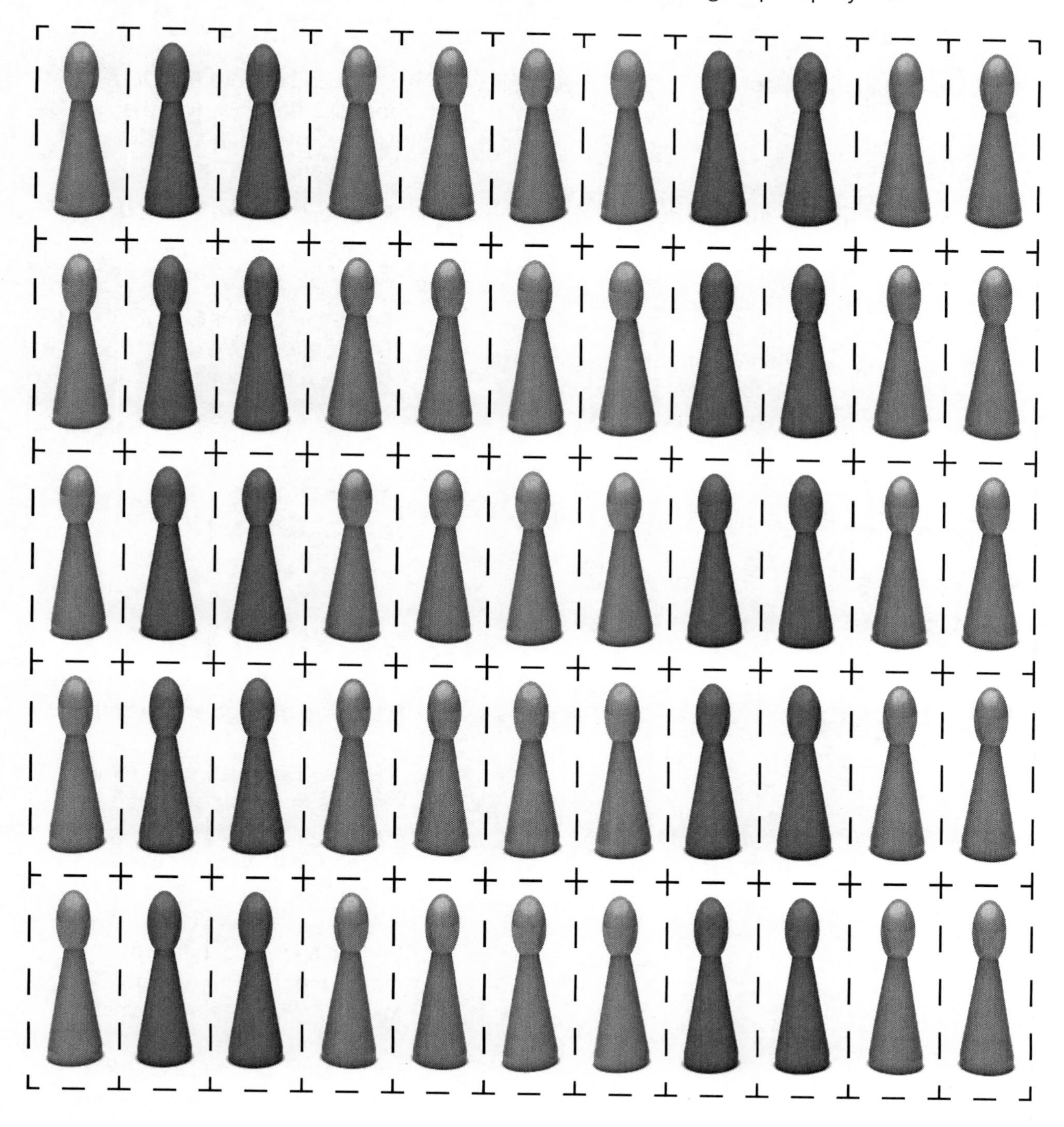

# Dive Into Distributions

## Domain

Statistics and Probability

## Standard

Giving quantitative measures of center (median and/or mean) and variability (interquartile range and/or mean absolute deviation), as well as describing any overall pattern and any striking deviations from the overall pattern with reference to the context in which the data were gathered.

## Number of Players

2 Players

## Materials

- *Dive Into Distributions Number Cards* (page 120)
- *Dive Into Distributions Recording Sheet* (page 121)
- paper bags
- scratch paper

## GET PREPARED

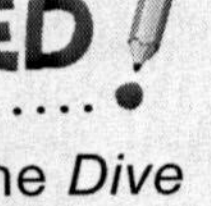

- Copy and cut out the *Dive Into Distributions Number Cards* for each pair of players.
- Copy a *Dive Into Distributions Recording Sheet* for each player.
- Place the *Dive Into Distributions Number Cards* in a paper bag for each pair of players.
- Collect scratch paper for each player.

## Game Directions

1. Distribute materials to players.
2. To begin Round 1, each player reaches into the paper bag and selects six *Dive Into Distributions Number Cards* without looking.
3. Each player calculates the mean, median, and range for the set of data on scratch paper, and records the information on the *Dive Into Distributions Recording Sheet*.
4. One point is awarded for the higher mean, higher median, and greater range.

 

# Dive Into Distributions *(cont.)*

5. The 12 number cards are not returned to the bag and are placed in a discard pile.

6. For Round 2, players repeat steps 2 to 5.

7. Round 3 is played in the same fashion. All 36 number cards will now be used.

8. Once statistics have been calculated and points awarded, have players determine the overall winner of the three rounds. The player with more points is the winner.

# Dive Into Distributions

## Number Cards

**Directions:** Copy and cut out cards for each pair of players.

| | | | | | |
|---|---|---|---|---|---|
| 1 | 2 | 3 | 4 | 5 | 6 |
| 7 | 8 | 9 | 10 | 11 | 12 |
| 13 | 14 | 15 | 16 | 17 | 18 |
| 19 | 20 | 21 | 22 | 23 | 24 |
| 25 | 26 | 27 | 28 | 29 | 30 |
| 31 | 32 | 33 | 34 | 35 | 36 |

 

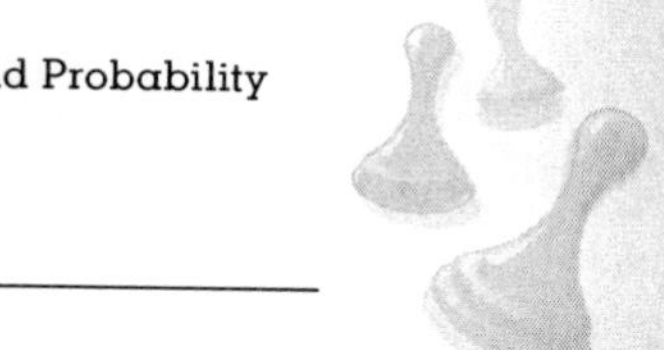

Name:______________________________ Date:______________

# Dive Into Distributions

## Recording Sheet

**Directions:** Select six numbers from the bag. Calculate the mean, median, and range for the numbers you selected. Take a point if you have the highest mean, median, or range. Play three rounds. Find the total number of points for the game. The player with the most points is the winner.

| Statistic | Round 1 | Round 2 | Round 3 |
|---|---|---|---|
| Mean | | | |
| Median | | | |
| Range | | | |
| Points | | | |

**Total points:** ______________

# Statistics Strikeout

## Domain

Statistics and Probability

## Standard

Giving quantitative measures of center (median and/or mean) and variability (interquartile range and/or mean absolute deviation), as well as describing any overall pattern and any striking deviations from the overall pattern with reference to the context in which the data were gathered.

## GET PREPARED!

- Copy and cut out a *Statistics Strikeout Game Board* and a *Data Grid* for each group of players.
- Copy and cut out the *Statistics Strikeout Game Markers* for each player.
- Collect a number cube for each group of players.

## Number of Players

2 to 3 Players

## Materials

- *Statistics Strikeout Game Board 1* (pages 124–125)
- *Statistics Strikeout Game Board 2* (pages 126–127)
- *Statistics Strikeout Data Grid* (page 128)
- *Statistics Strikeout Game Markers* (page 129)
- *Statistics Strikeout Answer Key* (page 134)
- number cubes

## Game Directions

1. Players take turns rolling a number cube. The player who rolls the lowest number is Player 1. Players take turns going in a clockwise order.
2. Players place their markers on start of the *Statistics Strikeout Game Board*.
3. Player 1 tosses the number cube and moves forward the indicated number of spaces.

# Statistics Strikeout *(cont.)*

4. The player locates the space on the *Statistics Strikeout Data Grid* corresponding to the ordered pair on the game board.

5. The player finds the indicated statistic for the data.

6. If correct, the player keeps their *Statistics Strikeout Game Marker* on the space.

7. If incorrect, the player moves back to the previous location.

8. Steps 3 through 7 are repeated by the other players.

9. The winner is the first player to reach the "Finish" space on the game board.

10. Answers can be checked using the *Statistics Strikeout Answer Key*.

# Statistics Strikeout

## Game Board 1

**Directions:** Copy and cut out the game board. Tape it to the game board on page 125.

| | | |
|---|---|---|
| **Range**<br>**(1, 3)** | **Median**<br>**(2, 2)** | **Mean**<br>**(3, 2)** |
| **Mode**<br>**(5, 6)** | | |
| **Median**<br>**(2, 6)** | | |
| **Mean**<br>**(4, 1)** | | |
| **↑**<br>**Start**<br>**Finish** | **Range**<br>**(6, 4)** | **Median**<br>**(4, 2)** |

 

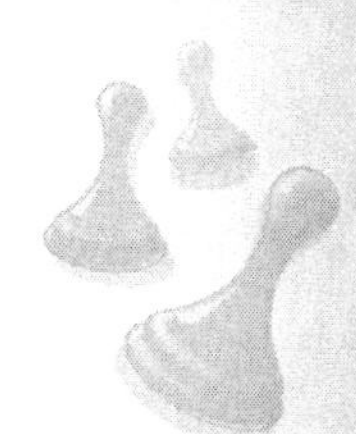

# Statistics Strikeout

## Game Board 1 *(cont.)*

tape here

| | | |
|---|---|---|
| **Range** (6, 6) | **Mean** (1, 2) | **Median** (3, 3) |
| | | **Mode** (5, 4) |
| | | **Range** (1, 6) |
| | | **Mean** (6, 5) |
| **Mean** (2, 5) | **Range** (5, 2) | **Median** (3, 6) |

# Statistics Strikeout

## Game Board 2

**Directions:** Copy and cut out the game board. Tape it to the game board on page 127.

| | | |
|---|---|---|
| **Range** (5, 1) | **Mean** (4, 3) | **Median** (1, 1) |
| **Mode** (3, 5) | Statistics S | |
| **Median** (6, 2) | | |
| **Mean** (1, 4) | | |
| ↑ **Start Finish** | **Range** (4, 6) | **Median** (2, 4) |

 

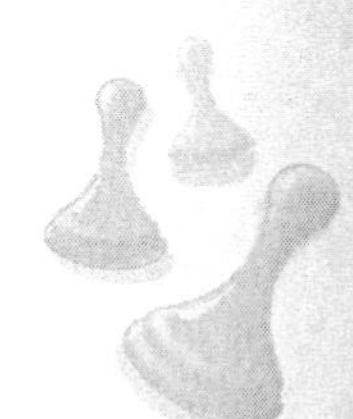

# Statistics Strikeout

## Game Board 2 *(cont.)*

tape here

| | | |
|---|---|---|
| **Range**<br>**(5, 5)** | **Mean**<br>**(3, 4)** | **Median**<br>**(5, 3)** |
| | | **Mode**<br>**(4, 5)** |
| | | **Range**<br>**(6, 1)** |
| | | **Mean**<br>**(2, 6)** |
| **Mean**<br>**(5, 2)** | **Range**<br>**(1, 5)** | **Median**<br>**(6, 3)** |

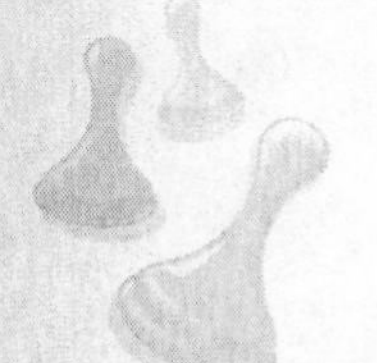

# Statistics Strikeout

## Data Grid

**Directions:** Use the numbers on this sheet as the data for the problem on the game board.

| | | | | | | |
|---|---|---|---|---|---|---|
| 6 | 15, 14, 18, 7, 3 | 15, 13, 7, 19 | 14, 6, 18, 5 | 16, 5, 14 | 10, 9, 12, 9 | 7, 3, 4 |
| 5 | 2, 10, 6 | 17, 10, 17 | 7, 8, 13, 7, 14 | 14, 20, 20, 9 | 20, 19, 15, 19 | 6, 20, 6, 3, 18 |
| 4 | 13, 2, 3, 4 | 1, 9, 7, 19 | 5, 1, 16, 17, 4 | 15, 13, 7 | 5, 17, 8, 6, 8 | 9, 8, 13 |
| 3 | 2, 10, 6 | 5, 4, 18 | 11, 6, 5, 16 | 13, 15, 8, 9 | 13, 17, 7, 19, 7 | 7, 18, 9, 2, 20 |
| 2 | 11, 4, 14, 18, 16 | 4, 15, 18, 1, 7 | 16, 14, 4, 5 | 18, 9, 7, 12 | 18, 10, 3 | 9, 13, 17 |
| 1 | 3, 7, 8 | 11, 9, 10 | 4, 10, 10, 13 | 16, 14, 12, 8 | 7, 12, 16, 19, 7 | 1, 11, 18, 12, 9 |
| | 1 | 2 | 3 | 4 | 5 | 6 |

# Statistics Strikeout

## Game Markers

**Directions:** Copy and cut out the game markers for each group of players.

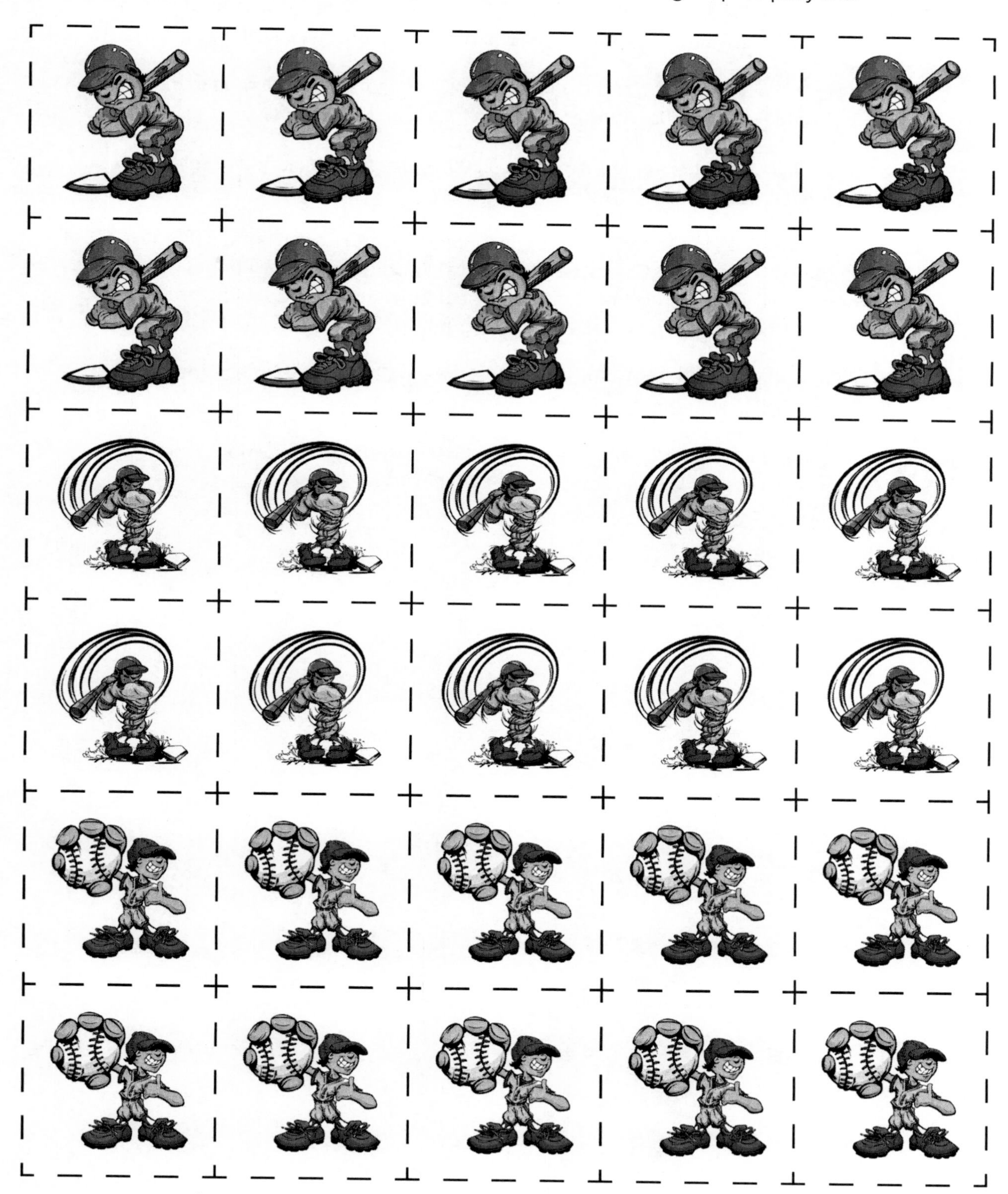

# References Cited

Burns, Marilyn. 2009. "Win-Win Math Games." *Instructor*. Reprinted March/April, http://www.mathsolutions.com/documents/winwin_mathgames.pdf.

Hull, Ted H., Ruth Harbin Miles, and Don S. Balka. 2013. *Math Games: Getting to the Core of Conceptual Understanding*. Huntington Beach, CA: Shell Education.

National Council of Teachers of Mathematics. 2000. *Principles and Standards for School Mathematics*. Reston, VA: NCTM.

National Governors Association Center for Best Practices, and Council of Chief State School Officers. 2010. "Common Core State Standards." Washington, DC: National Governors Association Center for Best Practices, Council of Chief State School Officers. Accessed September 23, 2013, http://corestandards.org/math.

National Research Council. 2001. "Adding It Up: Helping Children Learn Mathematics." Washington, DC: National Academy Press.

National Research Council. 2004. "Engaging Schools: Fostering High School Students' Motivation to Learn." Washington, DC: National Academy Press.

# Answer Key

**Roller Coaster Proportions (pages 20–24)**

| | | |
|---|---|---|
| 1. 2 cups of flour | 2. 8 hamburgers | 3. 9 robins |
| 4. 6 balls | 5. 5 tsp. of salt | 6. 8 quarts |
| 7. 10 centimeters | 8. 4 pencils | 9. 7 hours |
| 10. 1 girl(s) | 11. 3 green balls | 12. 9 liters |
| 13. 6 teachers | 14. 9 batteries | 15. 9 students |
| 16. 2 T-shirts | 17. 6 minutes | 18. 5 peaches |
| 19. 2 hours | 20. 1 egg | 21. 7 hours |
| 22. 4 houses | 23. 1 gallon(s) | 24. 7 costumes |
| 25. 10 pounds | 26. 6 cups of milk | 27. family of 3 |
| 28. 5 minutes | 29. 8 gallons | 30. 1 book |
| 31. 10 oranges | 32. 9 dogs | 33. 8 yellow blocks |
| 34. 5 feet | 35. $6 | 36. 3 blue |
| 37. 7 weeks | 38. 4 gallons | 39. 1 red shirt |

# Answer Key *(cont.)*

**Rocking Ratios (pages 26–33)**

| | | |
|---|---|---|
| $\frac{50}{50} = 1$ | $\frac{25}{74}$ | $\frac{25}{100} = \frac{1}{4}$ |
| $\frac{19}{10}$ | $\frac{10}{19}$ | $\frac{50}{33}$ |
| $\frac{25}{14}$ | $\frac{16}{10} = \frac{8}{5}$ | $\frac{13}{10}$ |
| $\frac{10}{20} = \frac{1}{2}$ | $\frac{10}{50} = \frac{1}{5}$ | $\frac{25}{50} = \frac{1}{2}$ |
| $\frac{40}{20} = \frac{2}{1}$ | $\frac{39}{51} = \frac{13}{17}$ | $\frac{30}{40} = \frac{3}{4}$ |
| $\frac{50}{25} = \frac{2}{1}$ | $\frac{14}{12} = \frac{7}{6}$ | $\frac{19}{10}$ |
| $\frac{19}{10}$ | $\frac{10}{50} = \frac{1}{5}$ | $\frac{13}{50}$ |
| $\frac{4}{9}$ | $\frac{7}{5}$ | $\frac{6}{10} = \frac{3}{5}$ |
| $\frac{9}{7}$ | $\frac{4}{5}$ | $\frac{2}{3}$ |
| $\frac{25}{19}$ | $\frac{14}{19}$ | $\frac{5}{5} = 1$ |
| $\frac{24}{60} = \frac{2}{5}$ | $\frac{7}{10}$ | $\frac{25}{50} = \frac{1}{2}$ |

# Answer Key *(cont.)*

**High Velocity Volume (pages 106–110)**

Start/Finish

V = 4 $m^3$

A = 16 $m^2$

V = 24 $ft^3$

A = 76 $m^2$

V = 16 $cm^3$

Empty

A = 66 $cm^2$

V = 27 $in^3$

A = 62 $in^2$

V = 6 $ft^3$

A = 54 $ft^2$

Empty

A = 40 $m^3$

Empty

A = 52 $ft^2$

V = 30 $ft^3$

A = 96 $in^2$

Start/Finish

# Answer Key *(cont.)*

**Statistics Strikeout (pages 122–129)**

| 6 | Mean 11.4<br>Median 14<br>No Mode<br>Range 15 | Mean 13.5<br>Median 14<br>No Mode<br>Range 12 | Mean 10.75<br>Median 10<br>No Mode<br>Range 13 | Mean 11.67<br>Median 14<br>No Mode<br>Range 11 | Mean 10<br>Median 9.5<br>Mode 9<br>Range 3 | Mean 4.67<br>Median 4<br>No Mode<br>Range 3 |
|---|---|---|---|---|---|---|
| 5 | Mean 6<br>Median 6<br>No Mode<br>Range 8 | Mean 14.67<br>Median 17<br>Mode 17<br>Range 7 | Mean 9.8<br>Median 8<br>Mode 7<br>Range 7 | Mean 15.75<br>Median 17<br>Mode 20<br>Range 11 | Mean 18.25<br>Median 19<br>Mode 19<br>Range 5 | Mean 10.6<br>Median 6<br>Mode 6<br>Range 15 |
| 4 | Mean 5.5<br>Median 3.5<br>No Mode<br>Range 9 | Mean 9<br>Median 8<br>No Mode<br>Range 18 | Mean 8.6<br>Median 5<br>No Mode<br>Range 16 | Mean 11.67<br>Median 13<br>No Mode<br>Range 8 | Mean 8.8<br>Median 8<br>Mode 8<br>Range 12 | Mean 10<br>Median 9<br>No Mode<br>Range 4 |
| 3 | Mean 6<br>Median 6<br>No Mode<br>Range 8 | Mean 9<br>Median 5<br>No Mode<br>Range 14 | Mean 9.5<br>Median 8.5<br>No Mode<br>Range 11 | Mean 11.25<br>Median 11<br>No Mode<br>Range 6 | Mean 12.6<br>Median 13<br>Mode 7<br>Range 12 | Mean 11.2<br>Median 9<br>No Mode<br>Range 18 |
| 2 | Mean 12.6<br>Median 14<br>No Mode<br>Range 14 | Mean 9<br>Median 7<br>No Mode<br>Range 17 | Mean 9.75<br>Median 9.5<br>No Mode<br>Range 12 | Mean 11.5<br>Median 10.5<br>No Mode<br>Range 11 | Mean 10.33<br>Median 10<br>No Mode<br>Range 15 | Mean 13<br>Median 13<br>No Mode<br>Range 8 |
| 1 | Mean 6<br>Median 7<br>No Mode<br>Range 5 | Mean 10<br>Median 10<br>No Mode<br>Range 2 | Mean 9.25<br>Median 10<br>Mode 10<br>Range 9 | Mean 12.5<br>Median 13<br>No Mode<br>Range 8 | Mean 12.2<br>Median 12<br>Mode 7<br>Range 12 | Mean 10.2<br>Median 11<br>No Mode<br>Range 17 |
| | 1 | 2 | 3 | 4 | 5 | 6 |

# Contents of the Digital Resource CD

| Student Resources | | |
|---|---|---|
| **Page(s)** | **Title** | **Filename** |
| 18 | Climb the Mountain Game Sheet | climbsheet.pdf |
| 19 | Climb the Mountain Game Markers | climbmarkers.pdf |
| 22–24 | Roller Coaster Proportions Cards | rollercards.pdf |
| 25 | Roller Coaster Proportions Recording Sheet | rollersheet.pdf |
| 28 | Rocking Ratios 1–100 Sheet | rocking100sheet.pdf |
| 29–32 | Rocking Ratios Cards | rockingcards.pdf |
| 33 | Rocking Ratios Recording Sheet | rockingrecsheet.pdf |
| 36 | Between Game Sheet | betweensheet.pdf |
| 37 | Between Game Sheet Markers | betweenmarkers.pdf |
| 40 | Dunking Decimals Game Sheet | dunkingsheet.pdf |
| 41 | Dunking Decimals Number Tiles | dunkingtiles.pdf |
| 44–45 | Mystery Multiples Game Board | mysteryboard.pdf |
| 46 | Mystery Multiples Game Markers | mysterymarkers.pdf |
| 49–51 | GCF for the Win Cards | gcfwincards.pdf |
| 52–53 | GCF for the Win Game Board | gcfwinboard.pdf |
| 54 | GCF for the Win Game Markers | gcfwinmarkers.pdf |
| 57 | Quotient Corral Game Sheet | quotientsheet.pdf |
| 62–65 | Rational Order Cards | rationalcards.pdf |
| 66 | Rational Order Number Line | rationalnumline.pdf |
| 69–71 | Variable Expression Cards | variablecards.pdf |
| 72–74 | Word Expression Cards | wordcards.pdf |
| 77–78 | Most Valuable 7 Game Board | valuable7board.pdf |
| 79–80 | Ordered Pair Cards | orderpaircards.pdf |
| 81 | Most Valuable 7 Game Markers | valuable7markers.pdf |
| 84–85 | Factored Concentration Cards | factorcards.pdf |
| 86–87 | Distributed Concentration Cards | distributecards.pdf |
| 90 | Equation Cards Set 1 | equationcards1.pdf |

# Contents of the Digital Resource CD *(cont.)*

| Student Resources | | |
|---|---|---|
| **Page(s)** | **Title** | **Filename** |
| 91 | Equation Cards Set 2 | equationcards2.pdf |
| 92–95 | Equation Bingo Game Board Set 1 | equationboard1.pdf |
| 96–99 | Equation Bingo Game Board Set 2 | equationboard2.pdf |
| 100 | Equation Bingo Markers | equationmarker.pdf |
| 103–105 | True or False: Inequalities Cards | truefalsecards.pdf |
| 108–109 | High Velocity Volume Game Board | velocityboard.pdf |
| 110 | High Velocity Volume Game Markers | velocitymarkers.pdf |
| 113 | Triangles Grid Sheet | trianglesgrid.pdf |
| 116 | Mean Wins Activity Sheet | meanwinssheet.pdf |
| 117 | Mean Wins Markers | meanwinsmarkers.pdf |
| 120 | Dive Into Distributions Number Cards | distributcards.pdf |
| 121 | Dive Into Distributions Recording Sheet | distributsheet.pdf |
| 124–125 | Statistics Strikeout Game Board 1 | statboard1.pdf |
| 126–127 | Statistics Strikeout Game Board 2 | statboard2.pdf |
| 128 | Statistics Strikeout Data Grid | statgrid.pdf |
| 129 | Statistics Strikeout Game Markers | statmarkers.pdf |

| Additional Resources | |
|---|---|
| **Title** | **Filename** |
| CCSS, WIDA, and TESOL | standards.pdf |